Python for 3D Printing

Using Python to enhance the power of OpenSCAD for 3D-modeling

by

John Clark Craig

A Books To Believe In Publication
All Rights Reserved

Copyright 2019 by John Clark Craig

No part of this book may be reproduced or transmitted in any form or by any means, electronic or mechanical, including photocopy, recording or by any information storage and retrieval system, without permission, in writing from the publisher.

Proudly Published in the USA by
Books To Believe In

publisher@bookstobelievein.com

Phone: (303) 794-8888

BooksToBelieveIn.com /python

First Edition: ISBN: 9781696881944

Dedication

For EJ,

*my inspiration and my partner
in all ways
and for always.*

Table of Contents

Introduction   9
    About Python
    About OpenSCAD
    OpenSCAD Limitations
    Python to the Rescue
    How Python was added to OpenSCAD
    How to Get Started
    Where to Get Python
    Where to Get OpenSCAD
    How to Learn from this Book

1 - Spheres	17
2 - Color	27
3 - Boxes	33
4 - Cylinders	39
5 - Tubes	43
6 - Cones	47
7 - Triangles	51
8 - Animation	57
9 - Polygons	63
10 - Polyhedrons	67
11 - Regular polygons	73
12 - Text	79
13 - Translate and Rotate	85

14 - Scale	95
15 - Resize	99
16 - Rotate extrude	103
17 - Spiral	111
18 - Hull	117
19 - Minkowski	125
20 - Mirror	133
21 - Projection	139
22 - Slice	145
23 - Offsets	151
24 - Difference, Union, & Intersection	157
25 - Assemblies	165
26 - Gears	169
27 - Mason bees	177
28 - Surface	181
29 - Platonic solids	187
Appendix A. openscad.py	193
About John Clark Craig	208
Other books by John Clark Craig	210

Python for 3D Printing

Introduction

About Python

Python is a well-known, easy-to-read, easy-to-write, and powerful programming language that has grown in popularity by leaps and bounds in recent years. If you are new to programming, check it out, as Python is a great choice for getting started. This book assumes you already have at least a minimal working knowledge of Python and isn't designed to walk you through its installation or teach you the basics of the language, although no overly complex concepts are used here. If you can create and run a simple "Hello World!" type of program using the IDLE editor that is included with Python, you should be all set.

About OpenSCAD

OpenSCAD is an amazing program for designing 3D parametric models for 3D printing, most often by creating industry standard STL files to command 3D printers. OpenSCAD is unlike most of the expensive software packages you perhaps have heard about or used. First, it's an open source, free program, instead of costing up to thousands of dollars. Another important feature is the way it works. Instead of interacting visually with a myriad of menus, sub-menus, buttons, complex settings, and parametrically driven object trees, OpenSCAD lets you drive the 3D modeling using a programming language all its own. It's a different way of thinking and creating, and once the few commands are understood, many people prefer it greatly over other programs.

Even though OpenSCAD is driven by text commands, the results are shown in an interactive 3D window, complete with panning, zooming, and rotating. This instant visual feedback makes creating your 3D objects a breeze.

OpenSCAD Limitations

So why bother with Python since OpenSCAD has its own programming language? The unique OpenSCAD syntax is like C, java, and several other programming languages, but it has some quirks I've never encountered before. Some of these quirks "drove me up the wall" while trying to get a handle on them.

As just one example, it's easy to store a number in a variable, like in any other programming language. But if you change the value of that variable later in the program, you'll likely be in for a surprise, as the value used throughout the entirety of your program will not be what you expected.

Also, the command sequence feels "backwards". For example, instead of creating a sphere at the origin, and then moving it to a new location or changing its color,

OpenSCAD has you stack up the modifying commands before you create the object to be modified. It works, but it feels different than the way most all other programming languages work.

Python to the Rescue

By creating an interpretive Python syntax layer that translates to OpenSCAD, these and several other unique issues are eliminated or minimized. You are required to learn and use only a simplified subset of standard Python syntax, which you possibly already know. (If you're new to Python, just know it's fantastically great for many other programming tasks as well.) Python enables more complex OpenSCAD models to be created in a shorter time because of its shorter learning curve, extendibility, flexibility, and the overriding of some limitations and quirks of the OpenSCAD language. In short, the combination of Python and OpenSCAD provides a very powerful and flexible system for creating 3D models in ways neither programming language alone can provide. All while having you learn a simple subset of the syntax of only one programming language, one of the most popular and powerful programming languages in the world.

How Python was added to OpenSCAD

Adding the Python interpretation and translation layer turned out to be easier than expected. My original attempt was to use an object-oriented approach, but in my opinion the syntax quickly became nearly as obtuse and problematic for programming as was OpenSCAD itself. I then tried a much simpler approach by manipulating simple strings to reconstruct the Python syntax into the equivalent OpenSCAD syntax, and this worked very well. The resulting code in openscad.py is amazingly concise and efficient, yet it still allows the redefining and restructuring I wanted, to create, in my opinion, a better approach to creating 3D models.

Appendix A provides the complete listing for the Python file named openscad.py that you simply need to add to your project's folder. Your Python script will import and use this code and you don't need to otherwise mess with it. Of course, if you wish to tweak the code, add to it, or whatever, feel free to do so! If you'd like a copy of this file, so you don't have to type it in and risk making any typo's, jot me a short note at john@openscadbook.com and I'll be glad to email it to you.

The way openscad.py works is relatively simple, compared to other language translation systems. The commands in the Python code are interpreted and turned into strings of commands that OpenSCAD expects. This sounds easy enough, but there are numerous details that needed to be overcome for this to work correctly. Many of OpenSCAD's commands are "backwards", in the sense that modifiers such as translate, rotate, and color are given before the object they operate on is defined. My goal was not to mimic OpenSCAD, but to create commands that felt more normal, and in line with the way most all other programming languages work, I figured out how to reverse the command order.

Other simplifications and changes were also made, such as to eliminate the confusing 2D constructs that OpenSCAD creates at zero thickness, even though on the screen the thickness is rendered at a fake 1 unit. Instead, all objects in this Python code are true 3D, and handling the implications for commands like rotate_extrude are handled in a much simpler way.

Try the commands as presented in the following chapters, and if anything works a little different than what you'd like, then feel free to make adjustments to the code in openscad.py.

How to Get Started

You can't really learn to ride a bicycle by reading a book or watching an online video. You need to go outside, get in the saddle, and try riding. With a little effort you'll quickly learn to get balanced and be a true biker. It's the same with Python and OpenSCAD. As you read these chapters, go ahead and give the code listings a try. Type them in, change the numbers, experiment freely, and see if you can predict what will happen if you make little changes here and there. You'll become a true 3D modeling guru much faster this way.

Above all, have some fun. There's something magical about typing a few short lines of commands and suddenly seeing a 3D object appear floating in space on your screen, ready to rotate, zoom, and pan to get a solid feel for its, well, solidness. It's very cool.

The first chapter presents a simple "Hello World!" program that simply creates a sphere in space. Be sure to read this recipe before jumping around to the others, because several key concepts on how to use the combination of Python and OpenSCAD easily and effectively are described in detail.

Where to get Python

If you haven't installed Python yet, go to
http://python.org/downloads
and download the most current Python 3 version for your operating system. Both Python and OpenSCAD run on Windows, Macs, and Linux systems, and they're both free and open source. If you need help getting up to speed with Python, check online or at your favorite bookstore.

Resources for learning are abundant.

Where to get OpenSCAD

If you haven't installed OpenSCAD yet, go to
http://www.openscad.org
and look for the download instructions for your operating system. You can use OpenSCAD in Windows, on your Mac, or in Linux. The Windows installation is very easy, very quick, and with very low impact on your system. Of course, a decently fast graphics card will help things run better and smoother.

OpenSCAD Cheat Sheet

Documentation for OpenSCAD is a little hodgepodge, and the quality varies a lot. My favorite starting point for researching features of the language is the official "cheat sheet" page that has links into the primary documentation for each command, key word, and system variable. Here's the link to that page:

http://www.openscad.org/cheatsheet/index.html

How to Learn from this Book

Learning by doing is what this book is all about. Each chapter adds a bit more knowledge, using working examples, to your understanding of Python for OpenSCAD. Each chapter is as short as possible, without being too short, to make it easier for you to type in, or copy and paste if desired, so you can interact with the code and try out new things. Skim through the recipes sequentially, and when you see something interesting, or something even the least bit mysterious to you, then play with that code for a while.

Each chapter can be used as a resource for using the various Python commands for creating and modifying 3D objects, but a lot of explanations are presented in a sequential way. I suggest working through many of the examples in a sequential order at first, and then use the book as a reference to refer to as needed. Most of the source code listings are very short, so it doesn't take a huge effort to try them out.

Why I created this book

I started my programming career with the BASIC programming language, eventually writing a bunch of popular books for Microsoft Press and O'Reilly Media. BASIC lets you try things out really quick, with the result that learning and absorbing all the syntax and language details happens fast and easy.

Python and OpenSCAD both have that same great nature to them. I think of Python as the new BASIC for today's world, and I'll never go back. Likewise, OpenSCAD is the new, better, cheaper, faster, user-interactive way to do 3D modeling. I've never once had an OpenSCAD model fail to create a watertight STL file on the first try, simply because of the unique way it works, and I'll never go back to the other, much more expensive and more error prone approach to 3D modeling.

Type in a few lines from these recipes, try to guess what will happen when you change some details, and THEN give it a try. I guarantee you'll learn very fast, while having a lot of fun, and in no time at all you'll be 3D printing some fun, productive, and amazing stuff.

Be the guru!

1- Sphere

Hello World!

Just about every modern programming language has "Hello World!" as its very simplest example program. This is useful to make sure you have the language up and running correctly, that you are entering text or commands correctly, and to basically get the basics out of the way.

As a first step, type the following short program into Python's IDLE editor and save it in a file named hello_world.py. Be sure to copy this book's special library file named openscad.py (see Appendix A) into the same folder where you save this hello_world.py file.

```python
# hello_world.py
import openscad as o

# Create the world
o.sphere(10)
earth = o.result()

# Send result to OpenSCAD
o.output(earth)
```

The Sphere() command has just the one diameter parameter, and a new sphere is always created with its center at the origin.

Run this Python code to create an output file that is automatically named hello_world.scad. Open this file in OpenSCAD to see the result and be sure to leave OpenSCAD running in preparation for the next step.

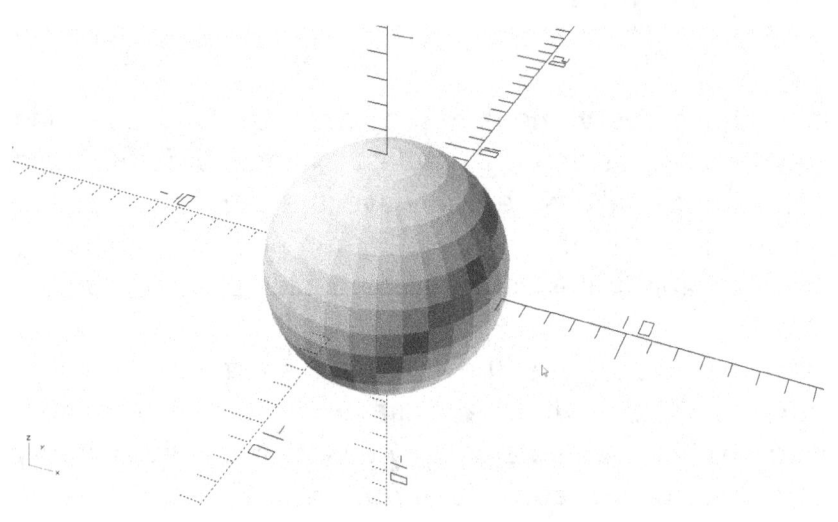

Using Python as an External Editor

If you leave OpenSCAD up and running, perhaps to the side of your IDLE window, you can make changes to your code and see the results instantly when you click Run from IDLE's menu. (Yes, you can use any other Python editor if desired, but in this book, I'll present all Python code using IDLE.)

You might need to do a one-time settings change to enable this automatic action. In the OpenSCAD window, from the "Design" menu click on "Automatic Reload and Preview". From then on, changes to your Python code automatically and nearly instantly show up in the OpenSCAD window as soon as you run your Python script. Keep both windows open side-by-side for maximum efficiency and enjoyment!

Try changing the sphere diameter to 15, instead of 10, click to run the Python code, and see the sphere instantly change size in the OpenSCAD window. Again, OpenSCAD detects changes to the hello_world.scad file and reloads it automatically. This feature makes our Python to OpenSCAD development environment seamless.

Hide the OpenSCAD Editor

You can close the source code editor window in the OpenSCAD environment to declutter your display. To do this, from OpenSCAD's "View" menu click on "Hide Editor".

You might want to study the OpenSCAD code that our Python program creates, in which case toggle this editor back on. In some cases, you might even want to tweak this OpenSCAD code, but that should be a very rare occurrence.

Here's an example of the OpenSCAD code that creates our world, in the shape of a sphere. In this case the source code is short and sweet, but in many programs, you'll find the Python code shorter and easier to understand, especially if you already know Python.

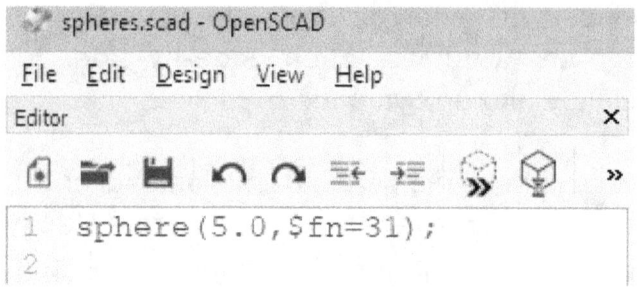

This program doesn't display the text "Hello World!" like most computer languages do, but this spherical globe looks like a world in space, so in a way it's much cooler.

If this is the first time you've created something in OpenSCAD, and you'd like to know more about how it works, you might want to study the online help documentation, or perhaps read the companion book to this one, titled *"OpenSCAD Cookbook"*.

How the code works

Let's take a close look at each line of our short Python program. Several parts of this code will be repeated throughout this book and understanding how it all fits together will be useful.

```
# hello_world.py
import openscad as o
```

The first line of code is a standard Python comment, where I've named the program's file for easy reference. You can type any notes or comments in comment lines. Comment lines are ignored when creating the output.

openscad.py file

The second line imports the special library file named openscad.py, where all the translation from Python to OpenSCAD takes place. As mentioned before, Appendix A provides the complete listing for this special file, but feel free to jot me a note at john@openscadbook.com and I'll be glad to email you a digital copy.

To use the openscad.py file, simply save it in the same folder as any program you create that calls it. You can use it without digging into the details of how it works in most cases. You will need to know the syntax of the various Python modeling commands it provides, but that's what this book is all about.

```
# Create the world
o.sphere(10)
earth = o.result()
```

After another comment line, the middle lines of this script create a sphere with a diameter of 10 units, and the results are saved in our variable named "earth".

About that "o." Prefix

You'll recognize all the new functions and variables called from the openscad.py file because they all use the "o." prefix, as declared in the import command at the start of the code.

Alternate use of the Library

You can optionally call the commands in the openscad.py library without the "o." prefix, by changing how the library is referenced in your code. I'll do this occasionally throughout this book as a reminder, although for clarity and to prevent any accidental namespace clashes, I usually use the "o." prefix as shown above. Here's an example of this alternate syntax.

```
# hello_world.py
from openscad import *

# Create the world
sphere(10)
earth = result()

# Send result to OpenSCAD
output(earth)
```

Programming pattern

The general pattern shown here is followed throughout this book, where one or more 3D objects are created and stored in one or more variables. These variables are output in a list (or as a single object as shown here) at the end of most code listings, as demonstrated in the last two lines of this program.

```
# Send result to OpenSCAD
o.output(earth)
```

The o.output command automatically creates an output file using the same base name as the current Python file, with an extension of ".scad" appended so the OpenSCAD program will be happy. In this case, our hello_world.py program outputs a file named hello_world.scad, located in the same folder.

Surface fragments

You probably noticed the sphere is comprised of many small, flat rectangles, and is not very smooth at all. The default resolution for creating curved surfaces is a compromise between speed and smoothness. It works well using the default setting, but it's easy to change this setting for smoother surfaces. A special global variable named o.fragments can be used to set the "number of fragments" for creating curved surfaces. There are other special variables in OpenSCAD that control the fragmentation if required, but setting o.fragments in our Python code will usually do the trick nicely for a sphere, or any of the other curved surfaces we'll create later on.

Try these commands to see a smoother sphere:

```python
# hello_world.py
import openscad as o

# Create a smoother world
o.fragments = 100
o.sphere(10)
earth = o.result()

# Send result to OpenSCAD
o.output(earth)
```

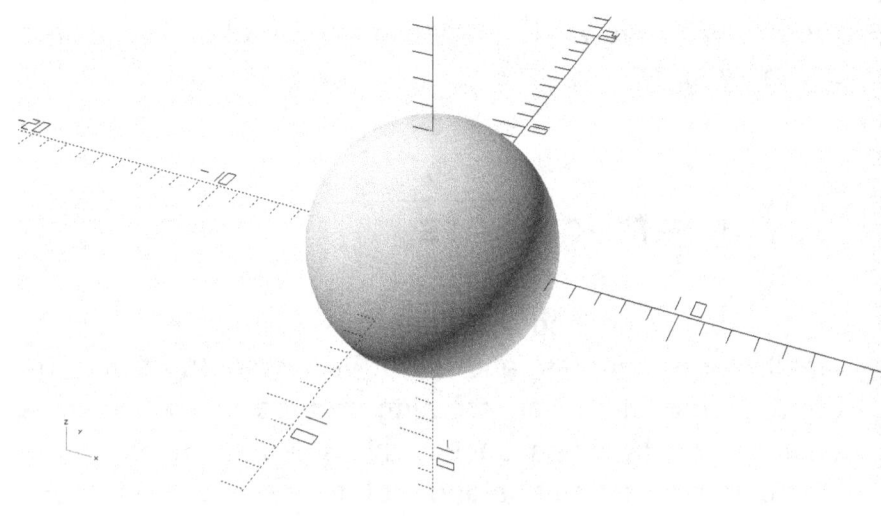

The default resolution, or fragmentation, is the same as setting the value of o.fragments to 30. Any number greater than 30 causes more rectangles to be drawn, creating a smoother appearance. Setting o.fragments to values less than 30 creates a rougher sphere surface. As an interesting experiment, try setting o.fragments to 4, 5, or other small numbers.

Experiment cautiously with larger numbers and be patient if nothing seems to be happening. Behind the scenes your computer is likely just working overtime to get all the geometry created. A value of 100 for o.fragments is usually all you need.

While developing objects it's a good idea to leave o.fragments at or near its default value to render quickly while retaining plenty of resolution to be able to see the "smooth" surfaces accurately. Also note that the goal is often to create an STL file, for handing off to a 3D printer, and larger settings for o.fragments will increase the size of the STL file, likely increasing the time it will take to print.

Python for 3D Printing

2 - Color

Two (or more) objects can be created, modified to change color, and then easily added to or subtracted from the output. The following code creates the Earth and Moon as two different sized spheres, colors each sphere, then translates the Moon out in space away from the Earth a short distance.

```python
# color.py
import openscad as o

# Create the earth
o.sphere(10)
o.color('aquamarine')
earth = o.result()

# Create its moon
o.sphere(3)
o.translate(20,10,10)
o.rgb(200,200,200)
moon = o.result()

# Send result to OpenSCAD
o.output([earth, moon])
```

A sphere is first created, colored with a colorful color name (see https://www.w3.org/TR/css-color-3 for more colors) and stored in a variable named earth.

Another smaller sphere is created at the origin, translated (moved) 20 units along the X axis, and 10 units along the Y and Z axes, colored using RGB numerical values, and saved in a variable name moon. Note that the red, green and blue (RGB) values are each in the standard range 0 to 255, with 0 as black and 255 as full intensity. In this example the three values of 200 combine to create a medium light shade of gray.

Alpha Channel

An optional fourth parameter can be added to either the color() or rgb() commands to set the alpha channel. Normally, alpha is set to 1.0 by default, and this causes objects to be completely opaque. An alpha value of 0.0 causes complete transparency. Try the following code to see the axes lines through the semi-transparent earth object, and the moon will be highly transparent.

```
# color.py
import openscad as o

# Create the earth
o.sphere(10)
o.color('aquamarine', 0.5)
earth = o.result()

# Create its moon
o.sphere(3)
o.translate(20,10,10)
o.rgb(200,200,200, 0.1)
moon = o.result()

# Send result to OpenSCAD
o.output([earth, moon])
```

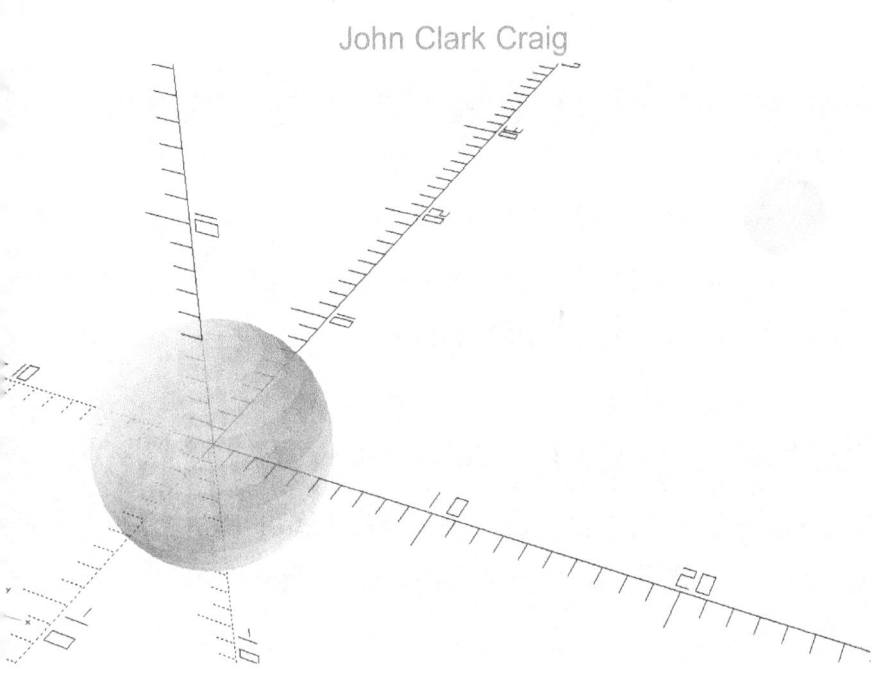

Output list

The last line of the code shows how the earth and moon objects are placed in a list (the square brackets creates the Python list) and are passed to o.output. An OpenSCAD file named color.scad is automatically created and you should load it into OpenSCAD for display.

It's informative to remove either the Earth or the Moon from the list passed to o.output. The only thing you need to change to try this is the last line in the code. When you create complex models comprised of many parts, the ability to easily add or remove objects from the output list lets you hide or show the various parts as desired.

Also note that the o.output command can be passed a single object, without the list brackets around it, and this simpler syntax feels more natural if just a single object is created.

Compound objects

Be aware that you may use a lot of commands to create a single object if you wish. As a simple example, the following code creates both the Earth and Moon spheres and stores both of these 3D constructions in a single variable named earth_and_moon.

```
# hello_worlds.py
import openscad as o

# Create the earth
o.sphere(10)

# Create its moon
o.sphere(3)
o.translate(20,10,10)

# Save both in one object
earth_and_moon = o.result()

# Send result to OpenSCAD
o.output(earth_and_moon)
```

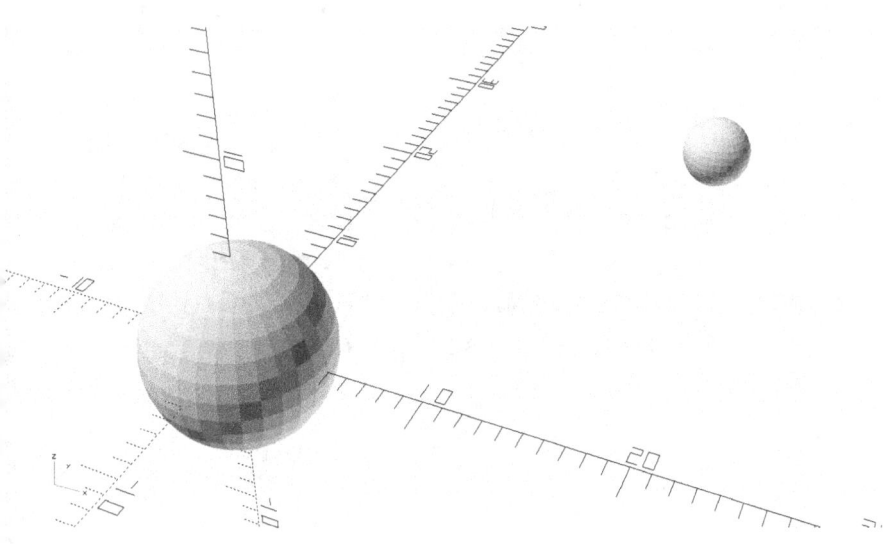

The way this works is that behind the scenes all modeling commands are accumulated as they are translated to OpenSCAD syntax, and whenever the results are stored in a variable using the o.result() command, the accumulated commands are all reset, ready for the next simple or complex object to be built. In general, I suggest building several simpler objects as you go, and putting them all in the o.output list at the end for finer granularity and control, but you can do it either way as you wish.

Viewing the Earth and Moon

Once these two objects appear on your screen in OpenSCAD, try rotating, panning, and zooming to see them from different angles and positions in space. Later in this book I'll show you how to animate the scene, so the Moon orbits automatically around the Earth. It's cool, but let's hold off on that until later.

The important concept to grasp here, in addition to the two ways to add color to your objects, is how an object is built up using any sequence of creation and modification commands, the result is stored in an appropriately named variable, and when ready, your various 3D objects are output in a Python list passed to o.output.

3 - Boxes

The box() command creates 3D boxes, cubes, and nearly flat squares of any desired size. For example, the following code creates a box 20 units long in the X direction, 10 units wide in the Y direction, and 5 units high in the Z direction.

```
# boxes.py
import openscad as o

# Create a box
o.box(20,10,5)

# Send result to OpenSCAD
o.output(o.result())
```

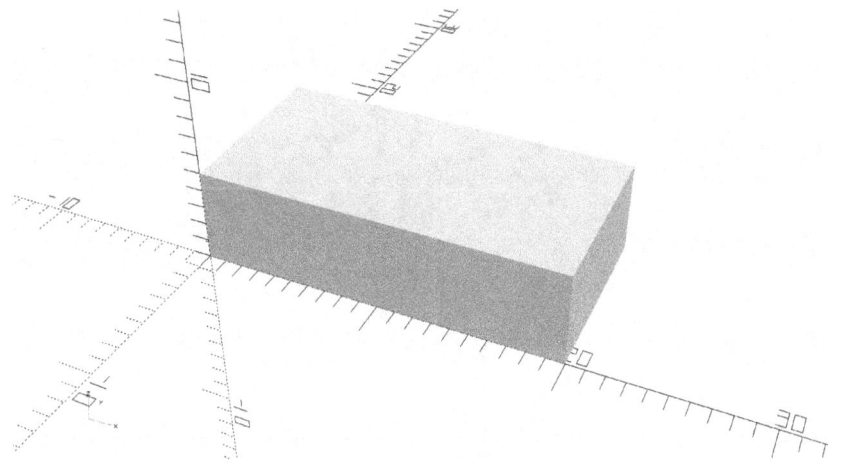

Before we go further, you perhaps noticed that this box was not saved in a named object variable before being sent to the output file. This demonstrates that you may optionally take the shortcut of saving the current results of your construction in one command, as shown in the last line of the code. I like saving my objects into variables with meaningful names, even in simple cases like this one, but it's good to know you can do it this shortcut way if desired.

A Cube is a Box

To create a cube, simply provide identical values for all three dimensions. This code creates a cube 10 units on each edge.

```
import openscad as o

# Create a 2D square
o.box(10,10,10)
cube = o.result()

# Send result to OpenSCAD
o.output(cube)
```

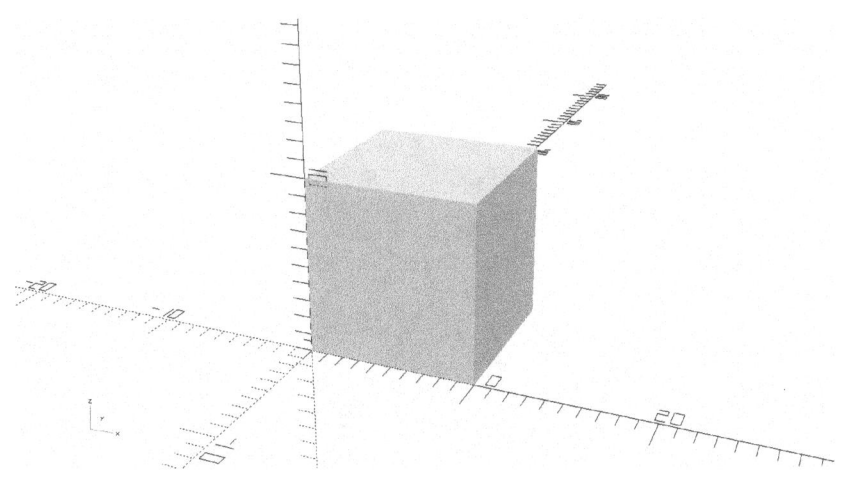

A Square is a Box

There are times when a very, very flat square is the thing to do. For example, the projection() and rotate_extrude() commands, which we'll get to in later chapters, often work best when an object that is basically a flat area is processed. OpenSCAD handles this whole subject in a more complex way, by providing separate commands for several 2D objects, whereas this Python approach simplifies both the code, and what you see on the screen, by simply letting you create extremely thin, but true 3D, objects.

Here's an example showing how to create a very, very flat square by providing the special constant value named o.tiny for the Z dimension.

Python for 3D Printing

```
import openscad as o

# Create a "square"
o.box(10,10,o.tiny)

# Send result to OpenSCAD
o.output(o.result())
```

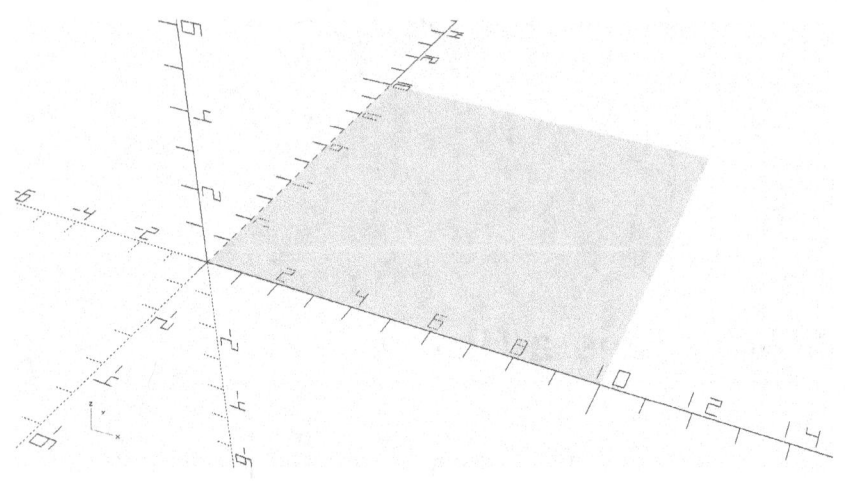

No matter how many times you zoom in on this square, you won't be able to see any thickness to it, although as far as OpenSCAD's engine is concerned, it does have a reasonable thickness. I've defined o.tiny as 1e-99, which means a whole bunch of zeros after the decimal point followed by a 1. This is truly a tiny number. OpenSCAD uses the C++ language math engine, and it allows even smaller numbers, so this value is a good compromise that works well to create a very flat "square" out of a rectangular solid.

Keep in mind that the contant o.tiny can be used other places, such as to create a flat circle from a very, very short cylinder. Also, even though the above code creates a square, you can use different values for X, and Y to create flat rectangles. It's even possible to create an "almost" perfectly one-dimensional line, by using o.tiny for two sides of a box, but you won't see it on the screen. Later, we'll see how the resize() command can change the shape of an object, and this is the command that I used to prove to myself the line I created actually did exist. We'll get to the resize() command later on.

A well centered existence

Perhaps you noticed that the spheres created in the first couple chapters were created with their centers at the origin, whereas all our boxes have appeared with one corner at the origin. The easy rule to remember is that objects with curved faces, such as spheres and cylinders, are centered at the origin, and straight-edged objects are created with a corner at the origin. Cylinders are a special case, centered in the X, Y plane around the origin, but the straight height goes up from there in the positive Z direction away from the origin, so even this composite figure comprised of curved and flat surfaces follows the same rule. We'll take a much closer look at the important cyl() command in the next chapter.

4 - Cylinders

Cylinders are important, as they can be tweaked to create rods, pipes, flat circles, cones, and other solid objects. To get started, here's how to create a basic cylinder shape with a diameter of 10 units and a height of 20.

```python
import openscad as o

# Create a cylinder
o.cyl(10,20)
my_cylinder = o.result()

# Send result to OpenSCAD
o.output(my_cylinder)
```

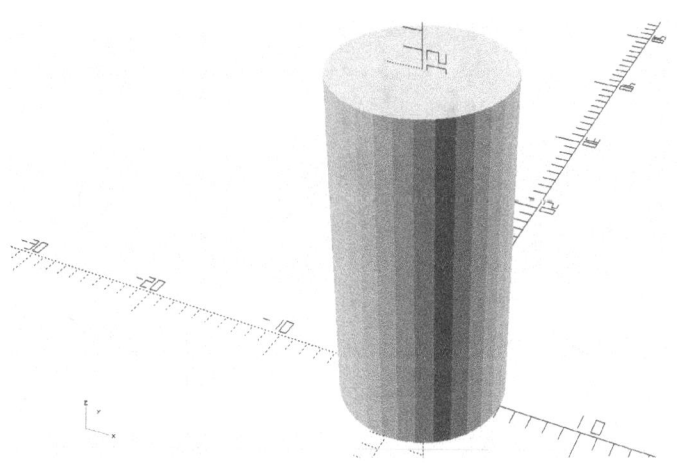

A Circle is a Cylinder

There are times when a flat circle is useful for constructing some more complex objects. The use of o.tiny lets us create a very short cylinder that is, for all intents and purposes, a very flat circle. This code, for example, creates a circle with a diameter of 10 units.

```
import openscad as o

# Create a "flat" circle
o.cyl(10,o.tiny)
circle = o.result()

# Send result to OpenSCAD
o.output(circle)
```

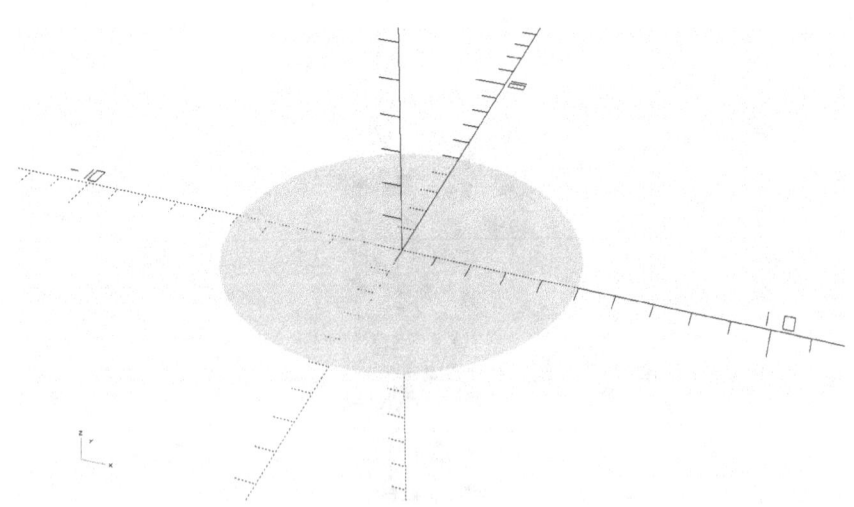

Again, just like when we created a "flat" square using the box() command, this circle does have a tiny amount of thickness to it, and it technically is a very squat cylinder, although you'll never be able to zoom in to see its height. When you need to start with a circle, this is the way to do it.

5 - Tubes

Pipes, tubes, and washer shapes are a common part of many 3D projects. The tube() command creates a hollow cylinder given inner and outer dimensions, plus a length.

In this example we'll create a Python function to convert inches to millimeters, and we'll use that to create a foot-long piece of standard 1" PVC pipe. There's no pressing reason to do the units conversion, other than to show an example of how to do so. However, many 3D printers default to mm units for all values found in an STL file, although this can be overridden since STL files have no built-in units at all.

Once converted to millimeters, the tube() command creates the hollow pipe in OpenSCAD, as shown.

Python for 3D Printing

```python
import openscad as o

def in_to_mm(inches):
    mm = inches * 25.4
    return mm

# Create a foot-long pipe of standard 1" PVC
length = in_to_mm(12)
outside = in_to_mm(1.315)
inside = in_to_mm(1)
o.tube(outside, inside, length)

# Send result to OpenSCAD
o.output(o.result())
```

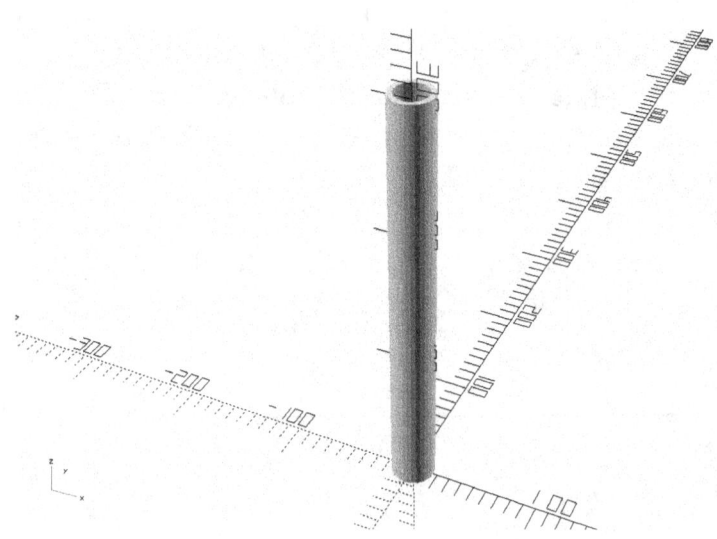

A Washer is a Tube

The tube command is also handy for creating washers, and similar squashed-down tube shapes. An example helps make this clear. The code below creates a standard washer for a ½ inch bolt, with measurements in millimeters that I found in a table on the Internet.

```python
import openscad as o

# Create a washer for 1/2" bolt
o.tube(34.9,14.3,2.6)

# Send result to OpenSCAD
o.output(o.result())
```

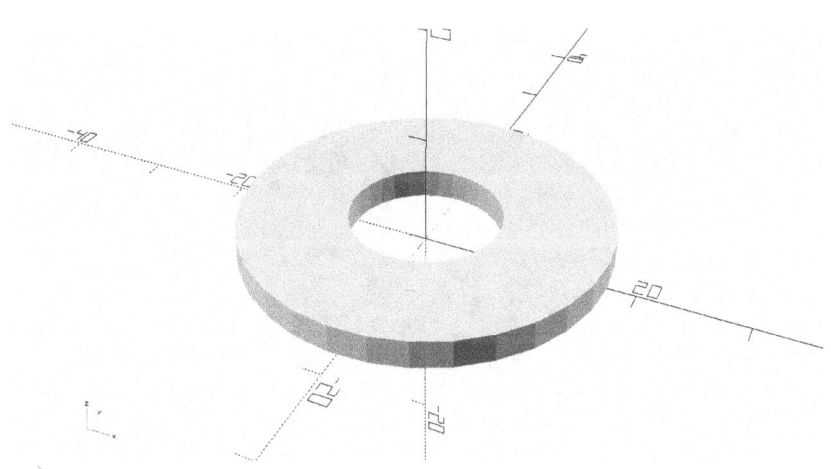

6 - Cones

The cone() command creates a cone given its base diameter and its height. The base of the cone rests in the X, Y plane, centered at the origin, and the height extends along the Z axis.

This example creates a cone with a base diameter of 10 units, and a height of 20.

```
import openscad as o

# Create a cone
o.cone(10,20)
cone = o.result()

# Send result to OpenSCAD
o.output(cone)
```

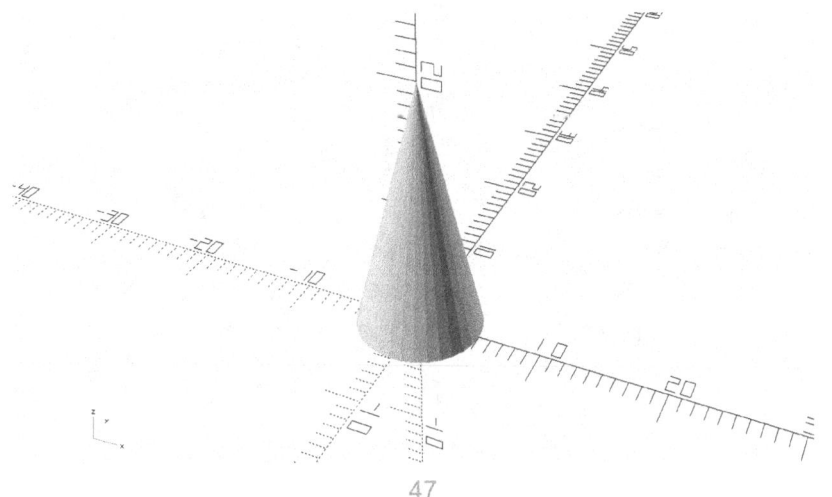

Just for fun, you might try setting the height to o.tiny, to create yet another very flat circle.

Hollow cone

It's easy to create a hollowed-out cone using the difference of two cone shapes. This is jumping ahead a bit, as the difference() command is explained in more detail later in the book, but the technique is so simple and informative that it's useful to present it here.

```
import openscad as o

# Create first cone
o.cone(10,20)
cone1 = o.result()

# Create second cone
o.cone(10,20)
o.translate(0,0,-1)
cone2 = o.result()

# Subtract the second cone
o.difference([cone1,cone2])

# Send result to OpenSCAD
o.output(o.result())
```

The two cones are identical in shape, but cone2 is translated down along the Z axis a bit, then subtracted in space from the cone1. Spin the displayed result around to look up from below to see the hollowed volume.

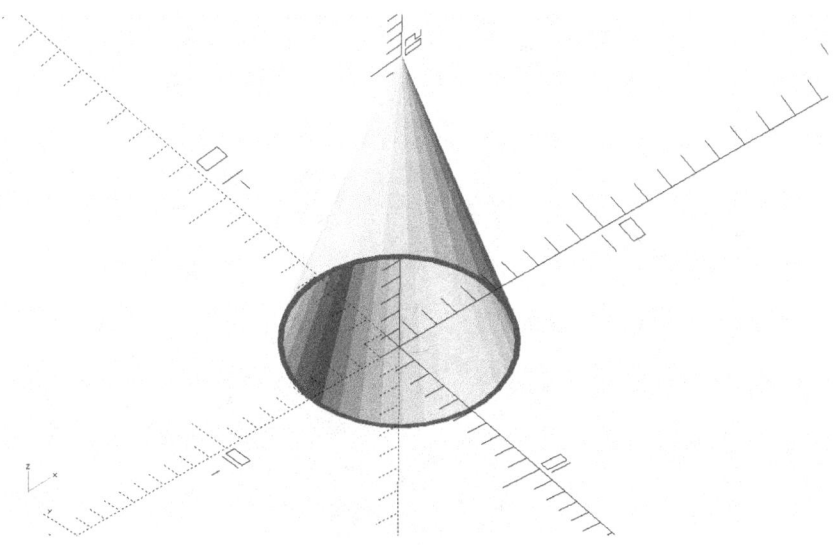

Both the translate() and difference() commands are explained later on, but it's fairly easy to grasp how they work in this case.

cone_truncated()

A closely related, but completely separate command is cone_truncated(). This command requires three parameters to define the truncated cone's base diameter, top face diameter, and its height.

The following example helps clarify what's going on. In this case the truncated cone has a base diameter of 20 units, a top face diameter of 10 units, and a height of 15.

Python for 3D Printing

```
import openscad as o

# Create truncated cone
o.cone_truncated(20,10,15)

# Send result to OpenSCAD
o.output(o.result())
```

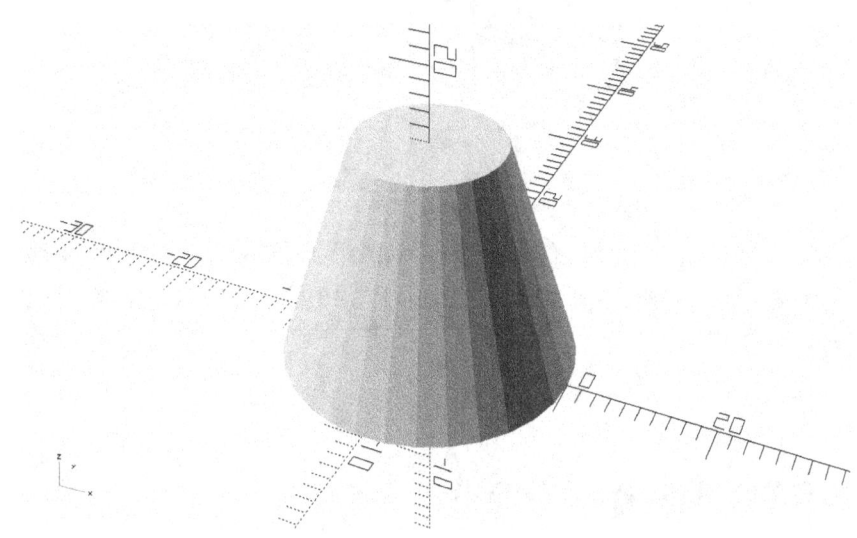

7 - Triangles

Triangles are the simplest straight-sided figures for creating areas, and they are also useful for constructing the faces of many 3D volumes.

Python lists, or collections of comma-separated values stored in square brackets, are perfect for creating X, Y points for each of the three corners of a triangle. The triangle() command expects three such points, plus a value for the triangle's thickness, or height in the Z direction. Here's an example of a triangle with a thickness of 1 unit.

Notice that the variable named triangle is completely separate from the function named o.triangle. If you use the "o." method to reference all functions and variables in openscad.py, this is a perfectly acceptable way to name your results.

```
import openscad as o

# Create a 2D triangle
p1 = [0,0]
p2 = [10,3]
p3 = [-5,20]
o.triangle(p1, p2, p3, 1)
triangle = o.result()
```

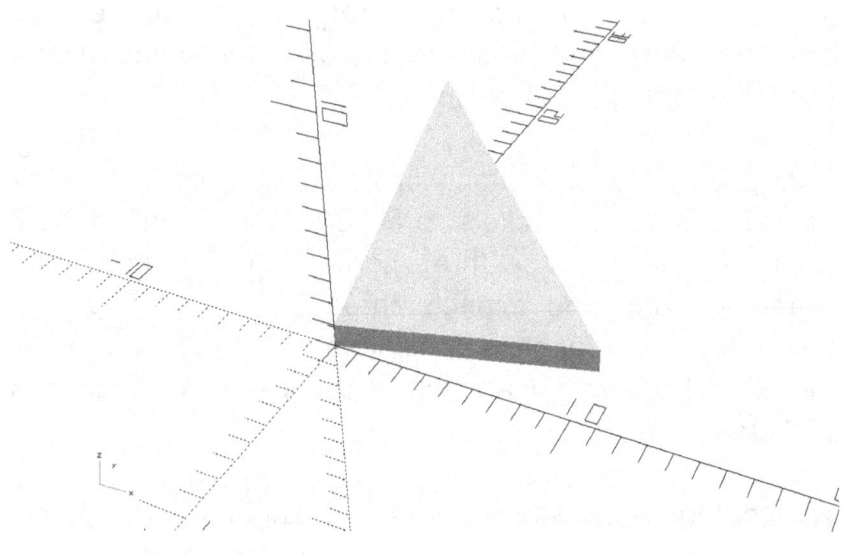

In the first example the three points are created and stored in separate variables, but the three points can just as easily be passed directly to the triangle command. This second example creates the same triangle, except that its height has been doubled to 2 units, and the output to OpenSCAD is more direct. This creates fewer Python variables, but it may be argued that the shorter code is created at the expense of some readability and clarity. Feel free to use whichever style of coding you prefer.

```
import openscad as o

# Create a short triangle
o.triangle([0,0],[10,3],[-5,20],2)

# Send result to OpenSCAD
o.output(o.result())
```

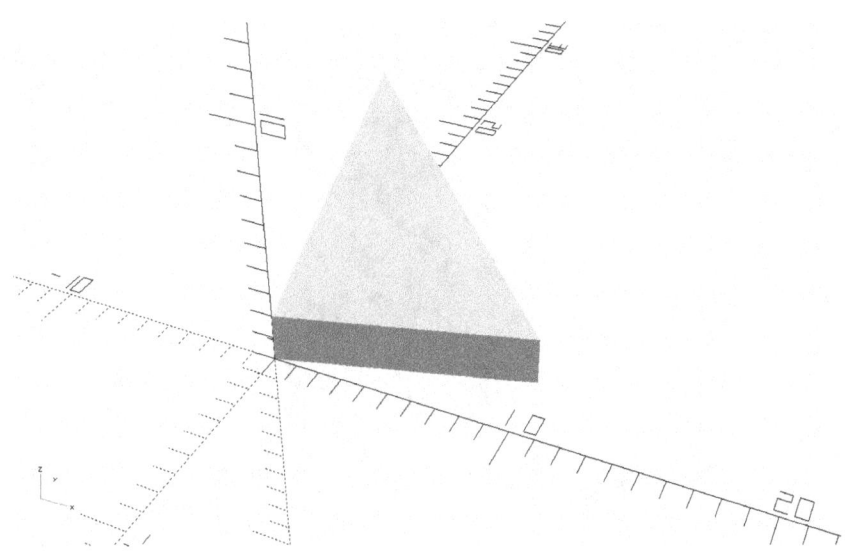

Flat triangles

There are times when a nearly 2D, very flat triangle provides a good starting point for a construction. Simply set the height parameter to o.tiny to create a very flat triangle. This next example creates a flat triangle, and just for fun it is colored pink. It's so thin that the Y axis marks show through.

```
import openscad as o

# Create a "flat" triangle
o.triangle([0,0],[10,3],[-5,20],o.tiny)
o.color("pink")

# Send result to OpenSCAD
o.output(o.result())
```

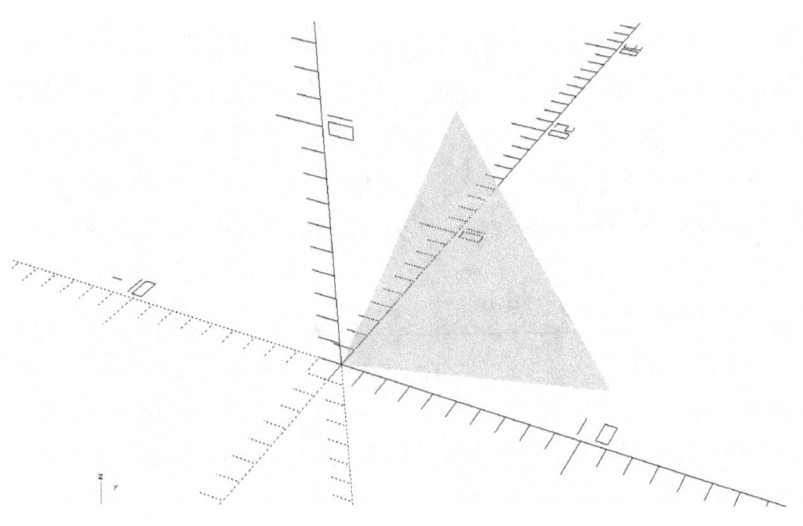

Finally, let's create a much thicker triangle, kind of like a chunk of cheese shape, and we'll use it to demonstrate several ways to form the three corner points. The first corner breaks down the construction of the point by first creating two floating point number variables, x1 and y1. These are combined into a list named p1 for the first point. The third point is created as x3 and y3, but they are combined directly into an unnamed point in the triangle() command. This example provides a little insight into how simple lists and points work in Python to provide the parameters required for the triangle() command, along with several other commands we'll explore later.

```python
import openscad as o

# Create the first point
x1 = -12
y1 = -4
p1 = [x1,y1]

# Set the second point
p2 = [10,-5]

# Define the third point
x3 = -5
y3 = 20

# Create a tall triangular solid
o.triangle(p1, p2, [x3,y3], 10)
triangle = o.result()

# Send result to OpenSCAD
o.output(triangle)
```

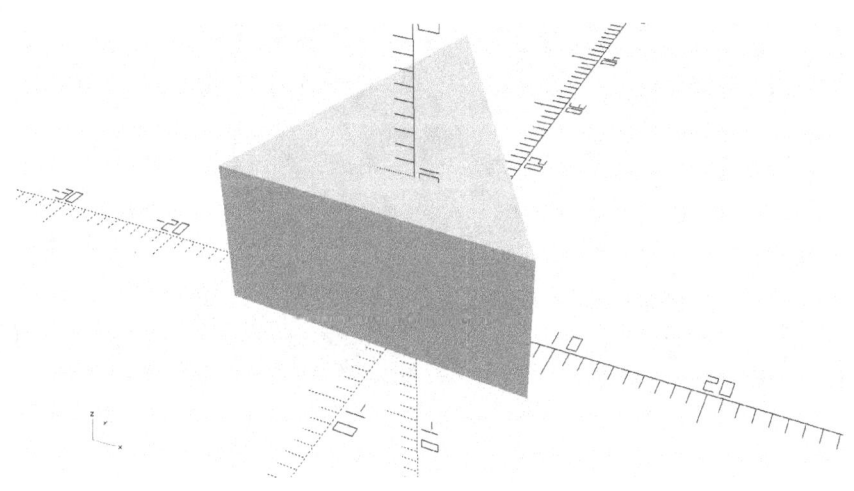

Python for 3D Printing

8 - Animation

The openscad.py library manipulates string data to efficiently translate Python based syntax to pure OpenSCAD. Because of this, there's a simple trick that allows very sophisticated animations to be created and played within the OpenSCAD environment.

To see how this works, let's return to our simple two-sphere program, where we created the Earth at the origin, and a smaller Moon sphere out in its orbit a little way from the origin. However, we'll pass string data containing a special OpenSCAD variable named $t, and this will open the door to some cool animation. Here's the code, then we'll take a closer look at how it works.

```python
# earth_moon_animation.py
# (Animate FPS: 30 Steps: 300)
from openscad import *

# Create the world
sphere(10)
color("green")
earth = result()

# Create its moon
x = '20*sin($t*360)'
y = '20*cos($t*360)'
z = '2*cos($t*360)'
sphere(3)
rgb(32,170,255)
translate(x,y,z)
moon = result()

# Send result to OpenSCAD
output([earth, moon])
```

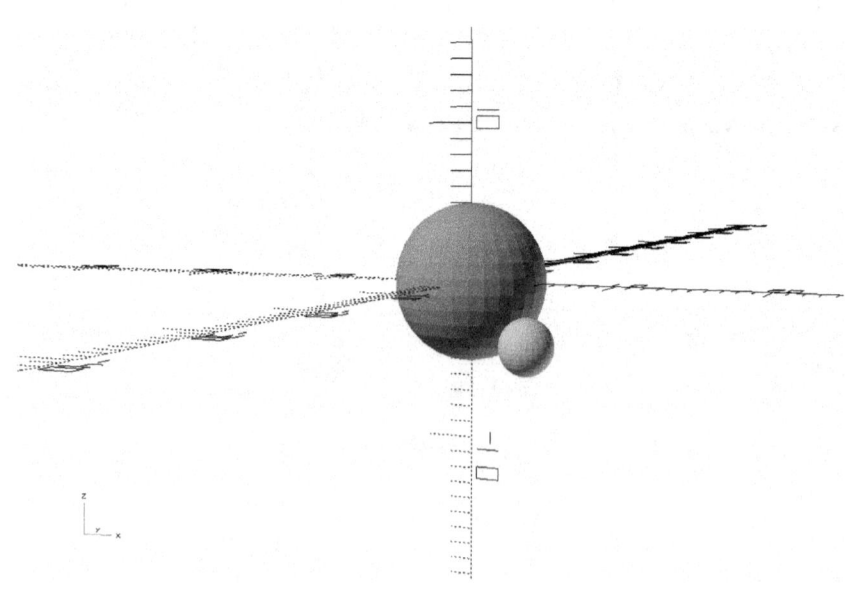

Three strings are created in this program, named z, y, and z. These string variables are passed to the translate() command, instead of the usual numerical values. It's important to note that the "code" within each string is pure OpenSCAD syntax, in this case providing calculations involving some trigonometric functions, and that mysterious $t variable.

Once animation is started in the OpenSCAD display, which I'll described how to do next, the value of $t automatically sweeps through values from 0.0 to 1.0 in a series of steps you control. At each step the image is updated, and any commands where $t has caused a new value will result in something new on the screen. In this case, the translated location of the Moon is mathematically calculated to sweep around the Earth as $t sweeps from zero to one.

Start the Action!

To start the animation, from OpenSCAD's View menu click on Animate. Three numerical input fields appear at the bottom of the display, labeled "Time:", "FPS:", and "Steps:" as shown here.

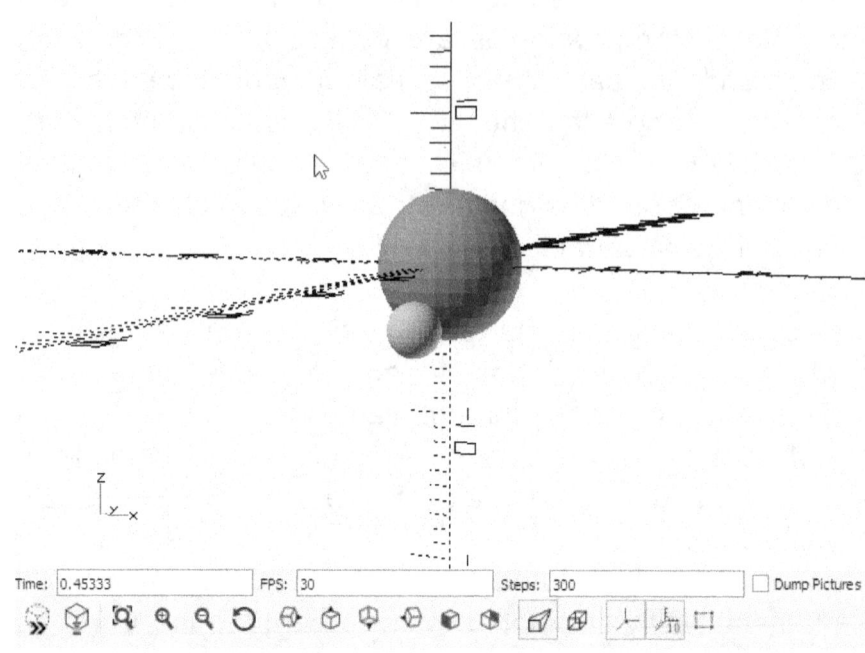

The time field is automatically filled during that animation, so there's no need for you to type anything there. The FPS field sets the frames per second. I suggest trying 30 to create a smooth update rate on your screen, but you may need to adjust this one way or the other based on your graphics card speed. The Steps field controls how many steps $t takes as it sweeps from 0.0 to 1.0. So, for example, if you also set Steps to 30, the Moon will spin around the Earth once a second, taking 30 jumpy steps to do so. Start with 300 steps and adjust this value as desired.

In the screengrab shown above, I caught the Moon as it was passing in front of the Earth when it was just a little short of halfway around its orbit, when $t reached 0.45333.

Animation in OpenSCAD is very flexible. $t can be used to sweep from color to color, by using it to calculate one or more of the RGB() command parameters. It can animate locations, dimensions, or basically anything where a calculated numerical variable can be used.

Animated GIFs for the web

Take another look at the screen grab showing the Moon in its animated orbit around the Earth. Just to the right of the Time field is a small check box labeled "Dump Pictures". Be careful when you select this, as a whole bunch of image files will suddenly start to show up in your project folder during the animation. But that can be a good thing!

There are online tools, and other graphics processing applications that allow a sequence of images to be collected and processed into a single *.gif file. The details of doing this is beyond the scope of this book, and it really does depend on the tools you select to create the GIF file. However, the results can be awesome for posting on the web, as most browsers will show the animation just fine.

Python for 3D Printing

9 - Polygons

Triangles can be used to piece together any complicated flat shape that has straight edges and where all the triangles are in the same plane. However, a polygon with any reasonable number of corner points can create the same resulting shape in a simpler way in many cases.

The polygon() command is passed a list of X,Y points, and a height. The height parameter can be set to o.tiny if a very flat polygon is wanted. The following example creates a simple 4-pointed polygon and sets its thickness, or height in the Z direction, to 2 units.

Note that each of the points is a list of two values, and these are combined into the list of points. This means the variable named points is a list of lists in Python. That sounds complicated, but if you just think of points as a list of however many points the polygon needs, it keeps the concept clear.

```
import openscad as o

# Create the polygon points
p1 = [10,0]
p2 = [-5,-5]
p3 = [-5,30]
p4 = [15,10]

# Create a list of these points
points = [p1,p2,p3,p4]

# Create a polygon
o.polygon(points,2)
polygon = o.result()

# Send result to OpenSCAD
o.output(polygon)
```

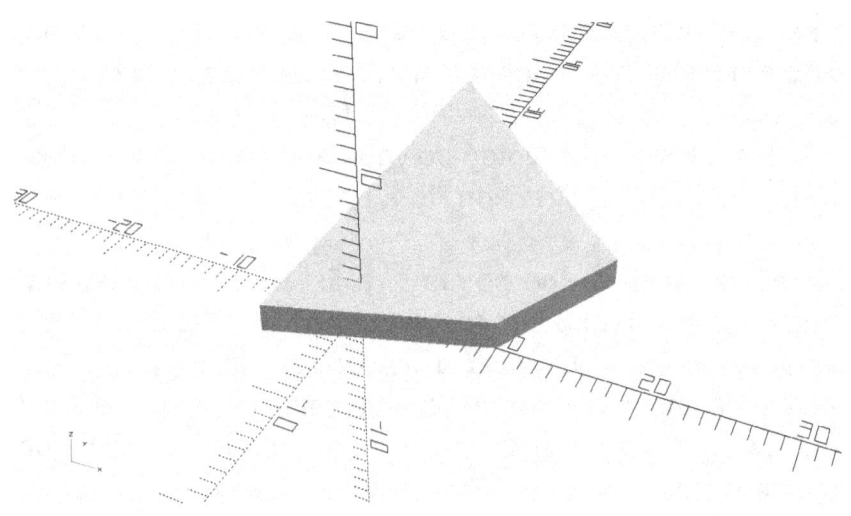

As mentioned before, it's easy to create a very, very flat, almost perfectly 2D polygon by setting the height parameter to o.tiny. Here's an example where the same polygon is created very flat.

```python
import openscad as o

# Create the polygon points
p1 = [10,0]
p2 = [-5,-5]
p3 = [-5,30]
p4 = [15,10]

# Create a list of these points
points = [p1,p2,p3,p4]

# Create a polygon
o.polygon(points,o.tiny)
polygon = o.result()

# Send result to OpenSCAD
o.output(polygon)
```

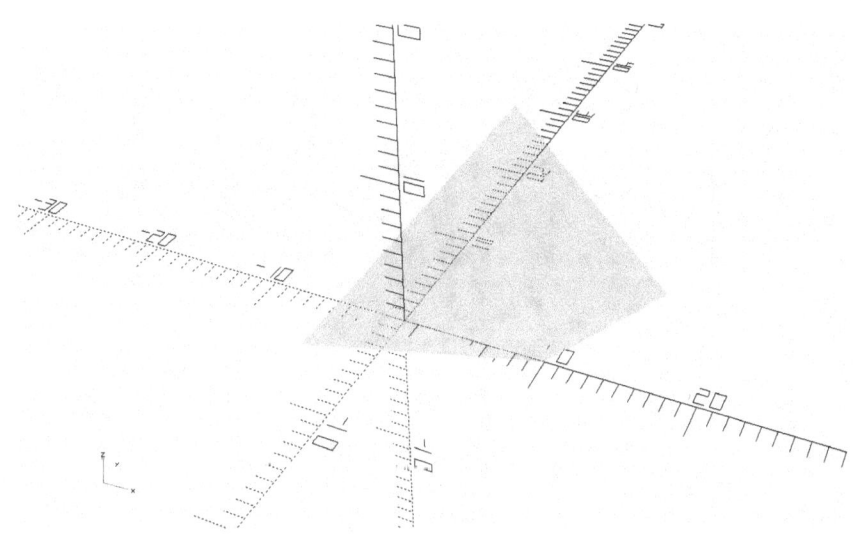

10 - Polyhedrons

A polyhedron is any volume in space that has flat surfaces and straight edges. A polyhedron requires at least four corner points, but it can have as many as you need. It's possible to construct complicated shapes, such as spiraling threads wrapping around a bolt, by creating one very complicated polyhedron, but I wouldn't attempt doing all the calculations for the points by hand. Python is great for setting up the looping and calculations required for this, and many other similar situations.

The following code creates a polyhedron with one corner at the origin, and the other corners along each of the three axes. The pattern for setting up the polyhedron command is the focus here, so we'll keep things simple.

The polyhedron command is passed two lists. The first list defines the coordinates for each corner point, where each point is a 3D list of X, Y, and Z coordinates.

The second list defines each face of the polyhedron, using a couple rules that you need to pay attention to. First, each list within the faces list contains indexes into the list of corner points in the order they occur, to define one face of the polyhedron. Each face will have its own list. Counting starts at zero, so the index numbers in these face lists will always be in the range zero to one less than the number of corner points.

The second rule to understand is that the order of the points in each face list is important, as the order defines which side of the face is on the inside of the polyhedron, and which side of the face is on the outside. The right-hand rule is useful here. If you mentally curl the fingers on your right hand to follow the points in a face list in the order presented, then your thumb should point towards the interior of the polyhedron.

I suggest studying the points and faces lists in the example code that follows until you grasp how this all works. The points list is straightforward, where each triplet of numbers tells you exactly in space where that point is located.

Our polyhedron has four faces, so there's four lists in the faces list, each defining the "walk-around" connecting the corner points to make each face.

```
import openscad as o

points = [[0,0,0],[10,0,0],[0,15,0],[0,0,20]]
faces = [[0,3,1],[1,3,2],[2,3,0],[0,1,2]]
o.polyhedron(points, faces)

# Send result to OpenSCAD
o.output()
```

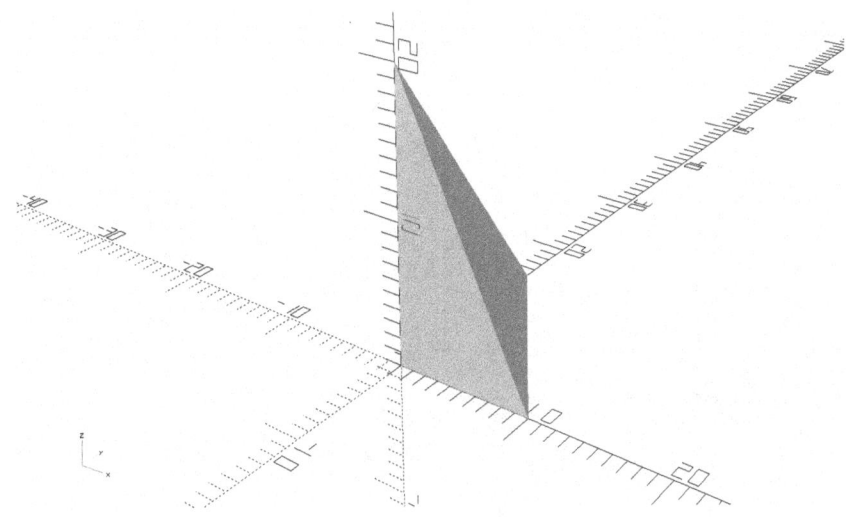

Unfolded prism

The following example is a rewrite of the example found in the online documentation for the polyhedron command in OpenSCAD. It's informative to compare how we can accomplish the same results using Python, with what I consider to be a simpler syntax.

Python for 3D Printing

```python
import openscad as o

def prism(l, w, h):

    points=[[0,0,0], [l,0,0], [l,w,0], [0,w,0], [0,w,h], [l,w,h]
    faces=[[0,1,2,3],[5,4,3,2],[0,4,5,1],[0,3,4],[5,2,1]]
    o.polyhedron(points,faces)

    # Preview unfolded
    z = 0.08
    separation = 2
    border = .2

    o.box(l,w,z)
    o.translate(0,w+separation,0)

    o.box(l,h,z)
    o.translate(0,w+separation+w+border,0)

    o.box(l,(w*w+h*h)**.5,z)
    o.translate(0,w+separation+w+border+h+border,0)

    o.triangle([0,0],[h,0],[0,(w*w+h*h)**.5],z)
    o.translate(l+border,w+separation+w+border+h+border,0)

    o.triangle([0,0],[-h,0],[0,(w*w+h*h)**.5],z)
    o.translate(-border,w+separation+w+border+h+border,0)
    # End of Preview unfolded

prism(10, 5, 3)
o.output(o.result())
```

John Clark Craig

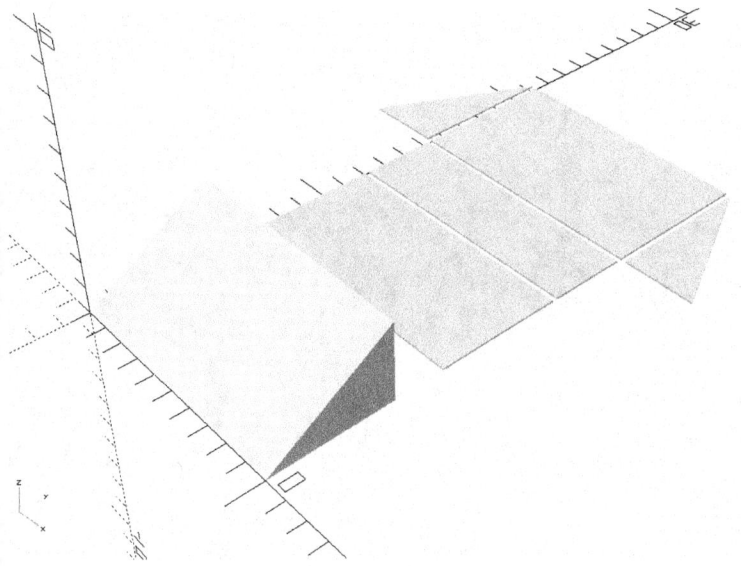

Python for 3D Printing

11 - Regular Polygons

Regular polygons have any number of sides, with all side lengths and interior angles equal. One common example of a regular polygon is the head of a bolt, where the number of sides is 6, and the thickness is a reasonable fraction of the "radius" of the hexagon shape.

The regular_polygon() command expects three parameters; the number of sides, the radial distance from the origin to a corner point, and the height or thickness. For example, this code creates a pentagon where the first point around the edge is at 10 units from the origin, and the height along the Z axis is 3 units.

```
import openscad as o

# Create a pentagon
o.regular_polygon(5,10,3)
pentagon = o.result()

# Send result to OpenSCAD
o.output(pentagon)
```

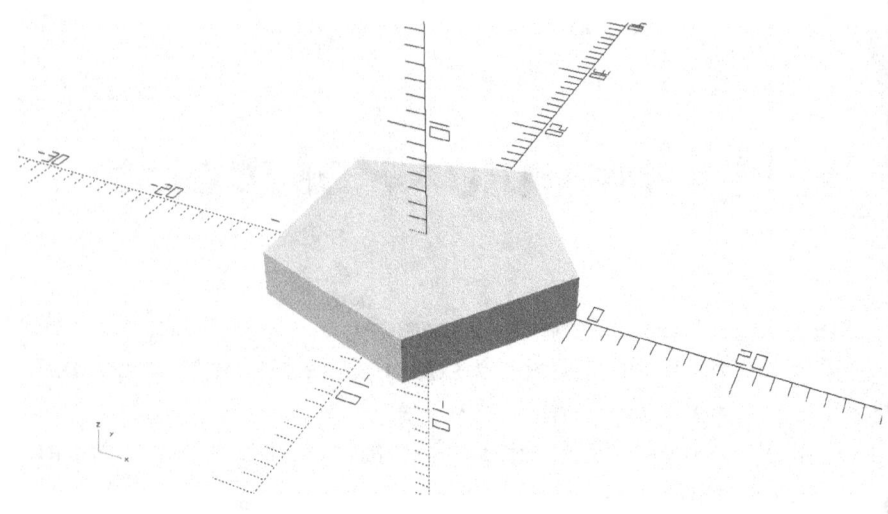

Here's a hexagon, or six sided regular polygon, created very flat by using a height of o.tiny.

```
import openscad as o

# Create a "flat" hexagon
o.regular_polygon(6,10,o.tiny)

# Send result to OpenSCAD
o.output(o.result())
```

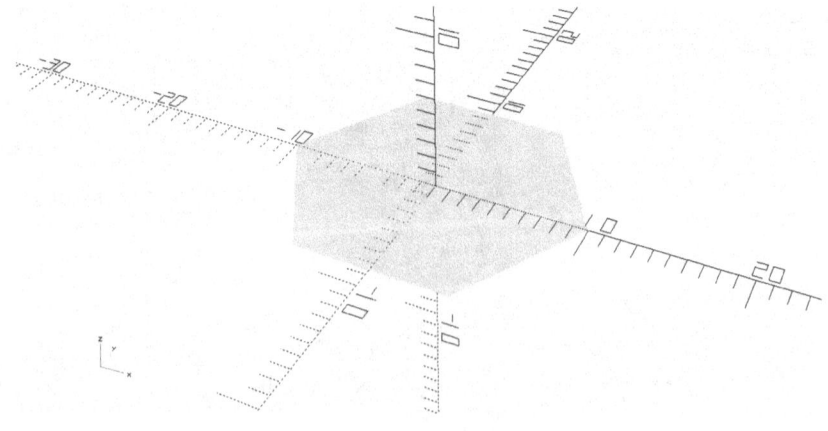

The next example creates a heptagon, or 7-sided polygon. In this case the height is set to 15 units to make a much taller volume.

```
import openscad as o

# Create a taller heptagon
o.regular_polygon(7,10,15)

# Send result to OpenSCAD
o.output(o.result())
```

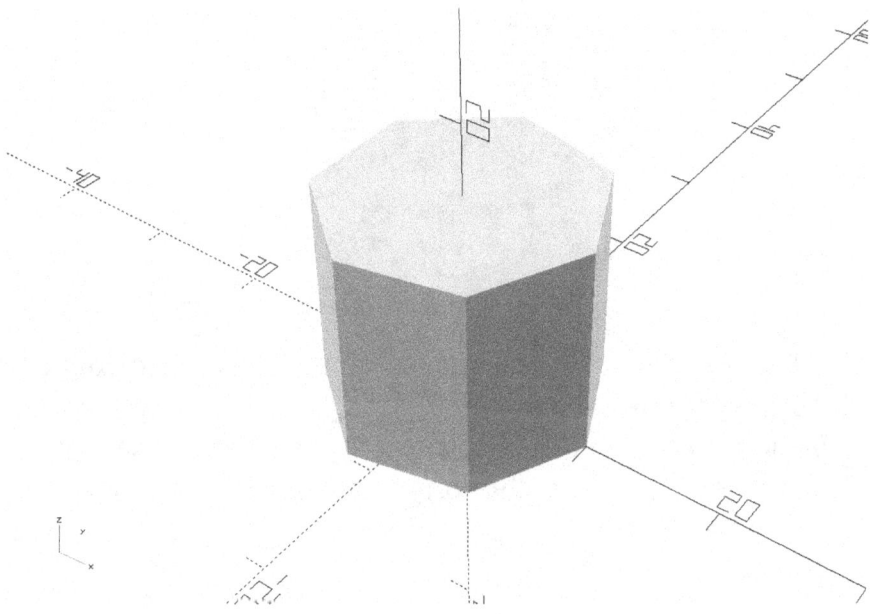

Regular polygon shapes abound in nature, but mostly in the man-made part of the world. Stop signs are octagons, so let's create one and color it red.

```
import openscad as o

# Create a red "stop sign" shape
o.regular_polygon(8,10,1)
o.rgb(255,0,0)

# Send result to OpenSCAD
o.output(o.result())
```

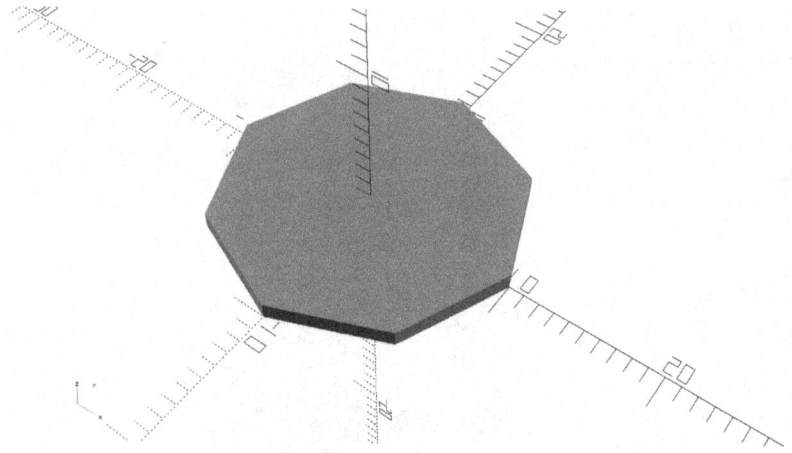

To be complete with all the common regular polygons, let's next create an equilateral triangle. Note that all these polygons start with one corner point on the X axis, at a given distance from the origin as set by the second parameter.

```
import openscad as o

# Create an equilateral triangle
o.regular_polygon(3,10,2)

# Send result to OpenSCAD
o.output(o.result())
```

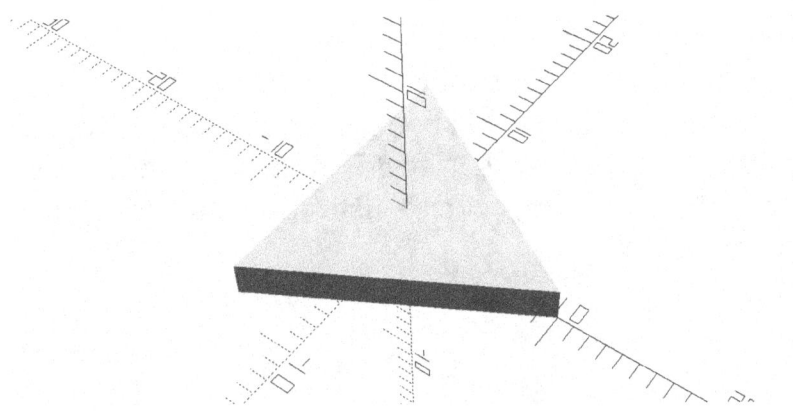

A four-sided regular polygon provides a new way to create a square. This square is rotated 45 degrees around the origin, but this might provide a handy starting point for some constructions.

```python
import openscad as o

# Create a square diamond shape
o.regular_polygon(4,10,1)

# Send result to OpenSCAD
o.output(o.result())
```

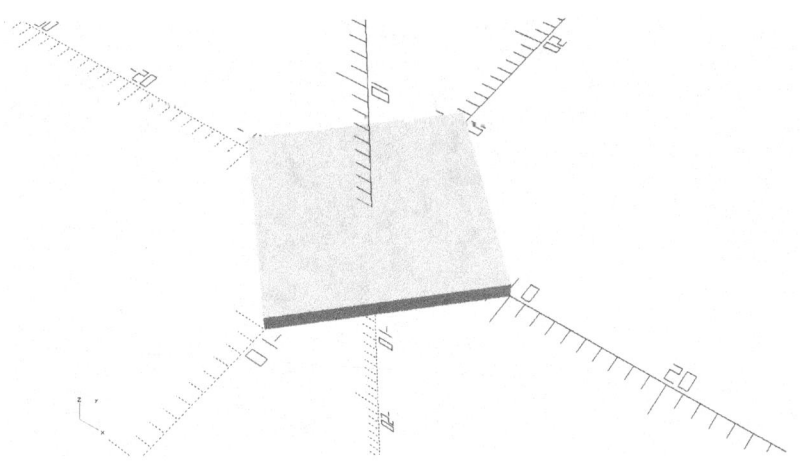

Finally, a regular_polygon with a lot of sides approaches the shape of a circle. This provides yet another way to create a circle, but I'd suggest using the cyl() command, as described earlier in this book to create your circles, as the resulting OpenSCAD code is much shorter. I've included the following "circle" just to let you know it can be done this way.

```
import openscad as o

# Create a "circle"
o.regular_polygon(100,10,1)

# Send result to OpenSCAD
o.output(o.result())
```

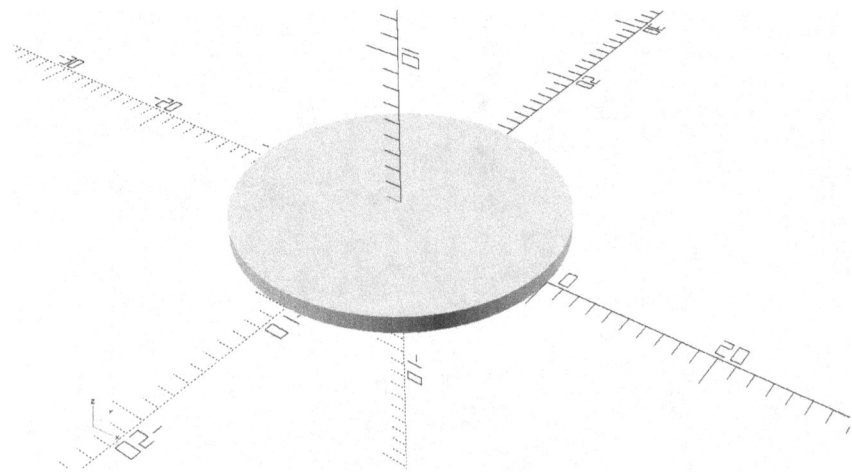

12 - Text

It's easy to add raised or recessed text to surfaces using the text() command. First, let's create some text in the form of solid figures, one for each letter of the text. The text() command needs, at a minimum, a string of text and our familiar height parameter to indicate how tall to make the letters along the Z axis.

There are several other optional parameters that will be described and shown in later examples. To help clarify exactly what is going on, all parameters other than the text string should be named. That's why this example code uses "height=5".

```
import openscad as o
o.fragments = 100

o.text("Python for OpenSCAD",height=5)

# Send result to OpenSCAD
o.output(o.result())
```

The o.fragments global variable was set to help smooth out the curved surfaces of the various text characters. Its use is optional of course.

Text on a surface

The next example creates a rectangular solid box and attaches the text to its surface. This creates a "rubber-stamp" like object that can be output as a single object to an STL file for 3D printing.

The box is colored "chocolate", and the text is translated up 10 units to mesh with the top surface of the box. The translate command is described in more detail in the next chapter.

```
import openscad as o
o.fragments = 100

o.box(150,20,10)
o.color("chocolate")

o.text("Python for OpenSCAD",height=3)
o.translate(7,6,10)

# Send result to OpenSCAD
o.output(o.result())
```

Recessed text

Sometimes it's better to hollow out, or recess text into a surface. This might be handy for imprinting a "Patent Pending" or similar message into a surface, without impacting that surface's interaction with other surfaces. The following example shows how this works.

You likely noticed the font and size parameters added to the text() command. There are several other parameters available to modify the text, and these will be described after this example.

Also, the difference() command begs some explanation. This will also be covered in another chapter coming soon, but for now just know that the difference() command subtracts one or more objects from the first one. Kind of like if the first object is Thanos, and all the rest in the list are reduced to dust and blown away. (No worries if that doesn't make sense, we'll demonstrate the difference() command soon enough.)

```python
import openscad as o
o.fragments = 100

o.box(150,20,10)
o.color("chocolate")
the_box = o.result()

o.text("Python for OpenSCAD",
       font="Liberation Sans:style=Bold Italic",
       size = 9, height = 5)
o.translate(7,6,9)
the_text = o.result()

o.difference([the_box,the_text])

# Send result to OpenSCAD
o.output(o.result())
```

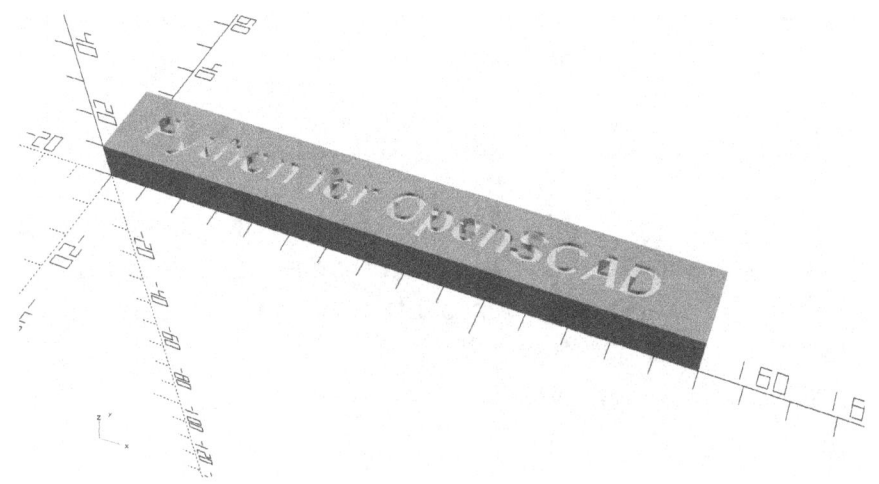

Here's a list of all the named parameters available when you use the text() command:

```
text="",
size=10,
font="Liberation Sans",
halign="left",
valign="baseline",
spacing=1,
direction="ltr",
language="en",
script="latin",
height=1,
```

The default value for each parameter is listed to the right of each equal sign above. For instance, the default text size is 10. Most of these parameters are usually ignored, but there are times when they come in handy. For instance, halign lets you center the text string at the origin or cause its right end to flush up against the origin.

To see a complete description of all these parameters, go to OpenSCAD's page that covers them all. Here's the link.

https://en.wikibooks.org/wiki/OpenSCAD_User_Manual/Text

The height parameter is not listed in the OpenSCAD documentation for the text command, as it was added when creating the Python syntax. All objects created by the openscad.py library are true 3D, requiring a height parameter in many cases.

13 - Translate and Rotate

The translate() command was previously used in several examples. It's time to explain this command in more detail.

The rotate() command works hand-in-hand with translate() to help move and position objects in space.

Let's take this step-by-step. First, we'll start with a simple rectangular solid created using the box() command.

```
import openscad as o

# Create a rectangular solid
o.box(2,5,9)

# Send result to OpenSCAD
o.output(o.result())
```

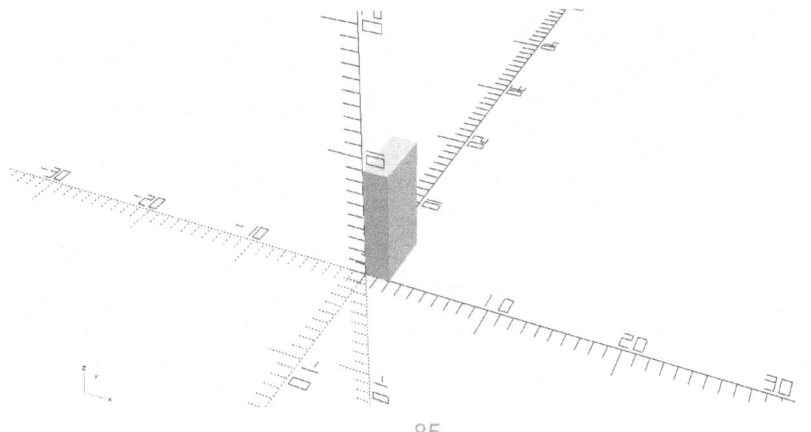

The three parameters in the translate() command cause the most recently created object to shift in position in the X, Y, and Z directions. Let's shift, or translate, our box 10 units in the X direction, then 5 units in Y, and nothing in the Z direction.

```python
import openscad as o

# Create a rectangular solid
o.box(2,5,9)

# Shift the position of our object
o.translate(10,5,0)

# Send result to OpenSCAD
o.output(o.result())
```

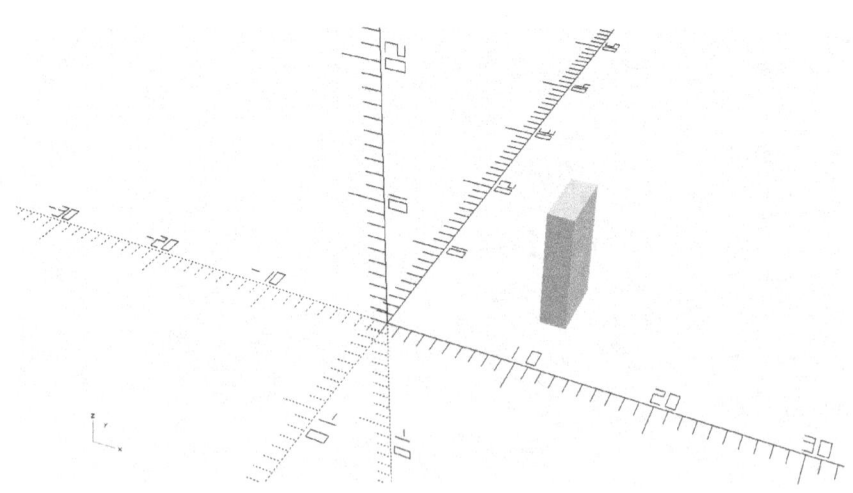

Now let's focus on the rotation() command. Using the same box, created with its corner at the origin, let's rotate around the X axis by 45 degrees. Recall the right-hand-rule? Always use it to picture the rotation direction. Point your right-hand thumb along the positive X axis and the curl of your fingers around your hand describes a positive rotation, in this case of 45 degrees.

```
import openscad as o

# Create a rectangular solid
o.box(2,5,9)

# Rotate around the X axis
o.rotate(45,0,0)

# Send result to OpenSCAD
o.output(o.result())
```

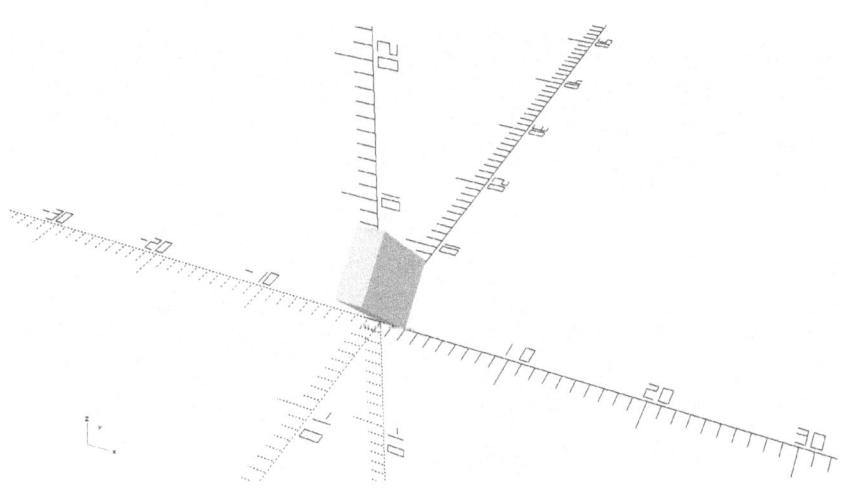

To rotate around the other axes, use something other than zero in the other rotate parameters. Note that the rotations always happen in the order of the parameters. That is, rotation around X happens first, then around Y, and finally around Z. The right-hand rule applies for each axis, where your thumb always points away from the origin.

If you want to rotate around the axes in a different order, you'll need to use more than one rotate command. For example, to rotate the box first around the Z axis by 45 degrees, then around the Y axis by -20 degrees, this is the way to do that.

```
import openscad as o

# Create a rectangular solid
o.box(2,5,9)

# Rotate around Z
o.rotate(0,0,45)

# Rotate around Y
o.rotate(0,-20,0)

# Send result to OpenSCAD
o.output(o.result())
```

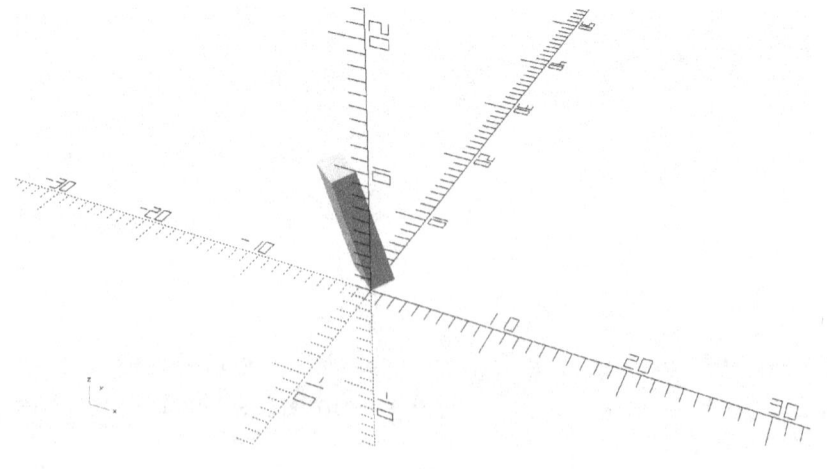

Order is important

As shown above, rotation order is important to understand to get the desired results. This is also very true when combining rotations and translations. If you keep in mind that all objects are first created near the origin, all rotations are around the three axes, and all translations are in the direction of the axes, you'll be able to plan for the transformations in space required.

Consider a rotation followed by a translation, as compared to the same translation first, followed by the same rotation. This example will help.

```
import openscad as o

# Rotate then translate
o.box(2,5,9)
o.rotate(0,0,90)
o.translate(10,0,0)
box1 = o.result()

# Translate then rotate
o.box(2,5,9)
o.translate(10,0,0)
o.rotate(0,0,90)
box2 = o.result()

# Send result to OpenSCAD
o.output([box1,box2])
```

As you can see, the boxes end up in entirely different places.

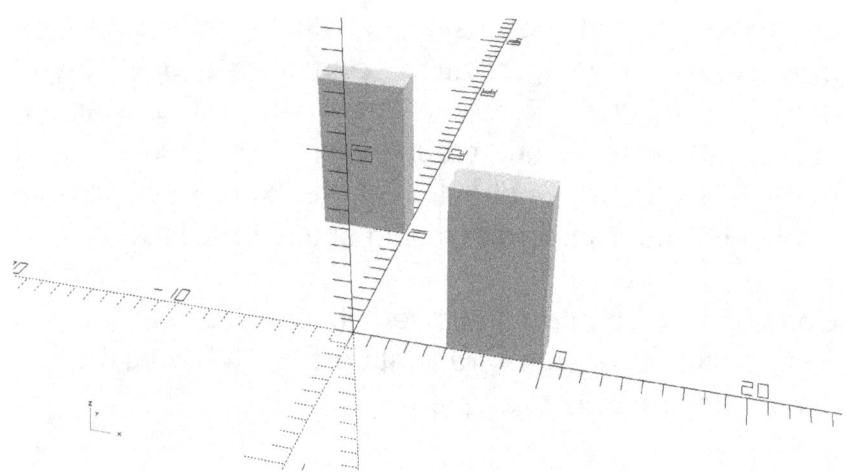

To reinforce some of the concepts about rotation order, the following example creates three long cylinders, each a different color, and rotates them in different ways.

```
import openscad as o

o.cyl(1,20)
o.rotate(45,45,45)
o.color("orange")

o.cyl(1,20)
o.rotate(0,0,45)
o.rotate(0,45,0)
o.rotate(45,0,0)
o.color("cyan")

o.cyl(1,20)
o.rotate(0,0,45)
o.color("chartreuse")

# Send result to OpenSCAD
o.output(o.result())
```

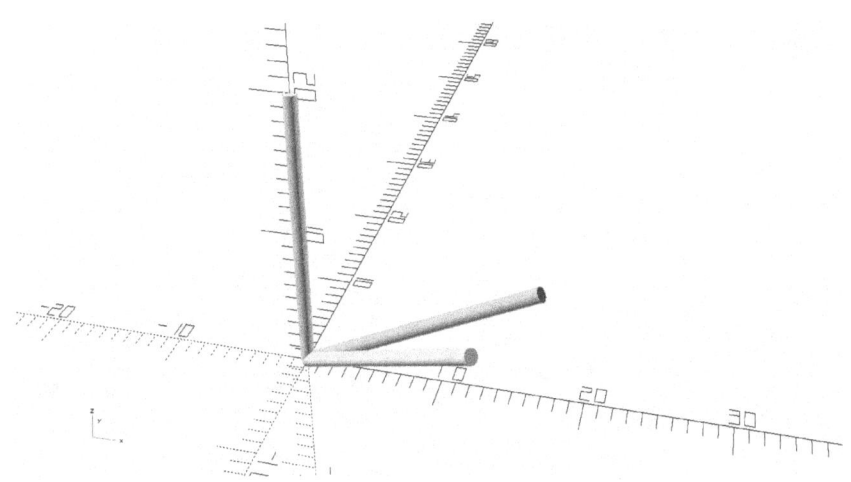

The first pole (cylinder) is rotated 45 degrees around each of the three axes, in X, Y, and Z order.

The cyan pole is rotated the same amount around each axis, but in exactly the opposite order. It ends up in a different orientation.

The green pole is rotated 45 degrees around Z. Notice that nothing happens in this case, as the cylinder was created along the Z axis, and rotation around that axis has no impact.

Disk art

As a final example, the following code creates 4 very flat and colorful circles, rotates each in space to a unique orientation, and the whole group is then combined and translated to a new point in space. The union() command is used to group the circles, and that command is explained in more detail in its own chapter.

Python for 3D Printing

```python
import openscad as o

# Create first flattish, red circle
o.cyl(15,o.tiny)
o.color("red")
disk1 = o.result()

# Rotate second blue circle around Y
o.cyl(12,o.tiny)
o.rotate(0,90,0)
o.color("blue")
disk2 = o.result()

# Rotate green circle around X
o.cyl(9,o.tiny)
o.rotate(90,0,0)
o.color("green")
disk3 = o.result()

# Rotate yellow circle around X, then Y, then Z
o.cyl(17,o.tiny)
o.rotate(45,10,-135)
o.color("yellow")
disk4 = o.result()

# Create a modifiable single object
o.union([disk1,disk2,disk3,disk4])

# Translate the whole construct
o.translate(15,10,5)

# Send result to OpenSCAD
o.output(o.result())
```

John Clark Craig

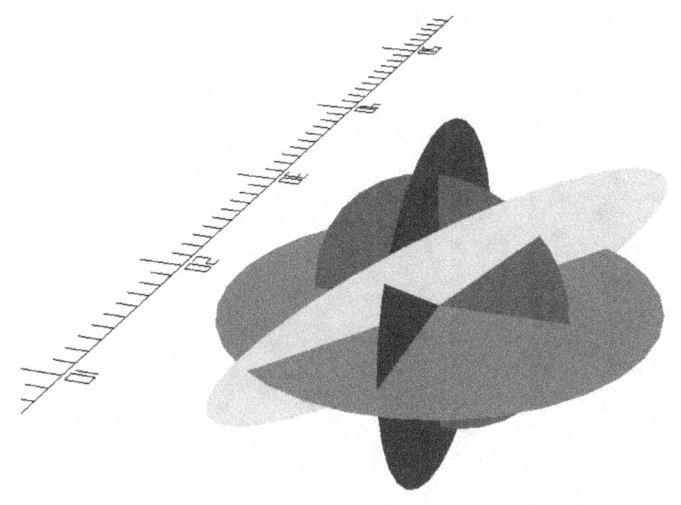

Python for 3D Printing

14 - Scale

The scale() command lets us stretch or shrink an object in each of the three axes directions. A value of 1 causes no change, less than 1 causes shrinkage, and greater than 1 causes stretching. The numerical value is simply multiplied by the current dimension to create the new value in the given direction.

```python
import openscad as o

# Create and scale a sphere
o.sphere(10)
o.scale(3,2,1/2)

# Send result to OpenSCAD
o.output(o.result())
```

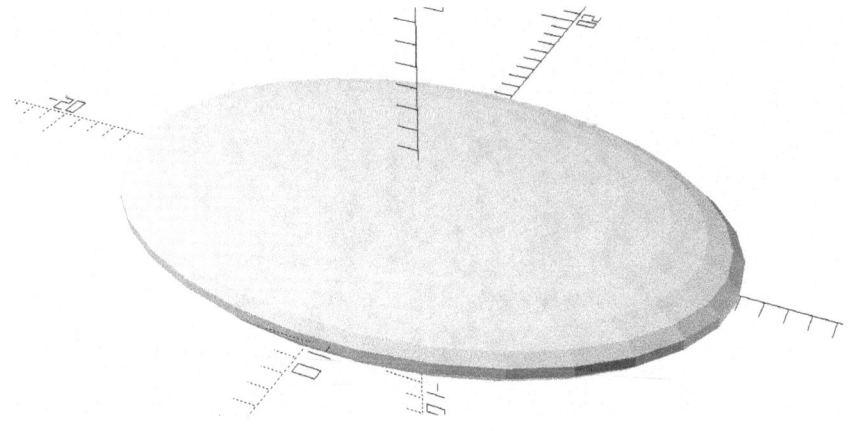

In this example a sphere with diameter 10 is created at the origin. The sphere is then stretched in the X direction by a factor of 3, in the Y direction by 2, and it is shrunk in the vertical Z direction by a factor of 0.5.

Culvert pipe

As another example, we can create a tube shape, then scale it to stretch it along one direction, to end up with something like those curved corrugated tubes placed under roadways to let creeks get through to the other side. Or something like that.

```
import openscad as o

# Create a tube
o.tube(10,9.8,20)
o.scale(2,1,1)

# Send result to OpenSCAD
o.output(o.result())
```

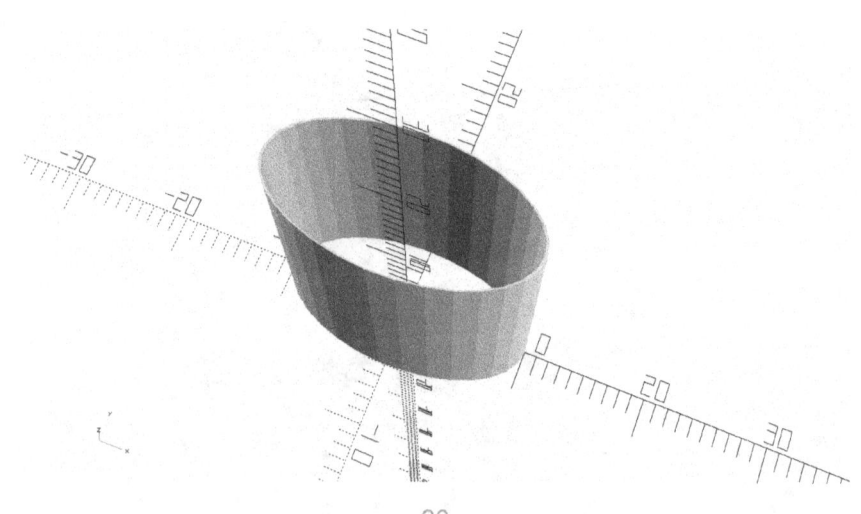

Be sure to use a value of 1 for any dimension where no stretching or shrinking is desired. It's easy to accidentally put a zero in there, and the results are ugly if you do this.

Python for 3D Printing

15 - Resize

The resize() command is similar in concept to the scale() command. However, instead of multiplying each dimension by a factor, the resize parameters define an exact size for the object in each of the three axes directions. In this example a sphere with a diameter of 1 unit is resized to 10 units width along the positive and negative X axis, to 5 units along Y, and 2 units high along Z.

```
import openscad as o

# Create and resize a sphere
o.sphere(1)
o.resize(10,5,2)

# Send result to OpenSCAD
o.output(o.result())
```

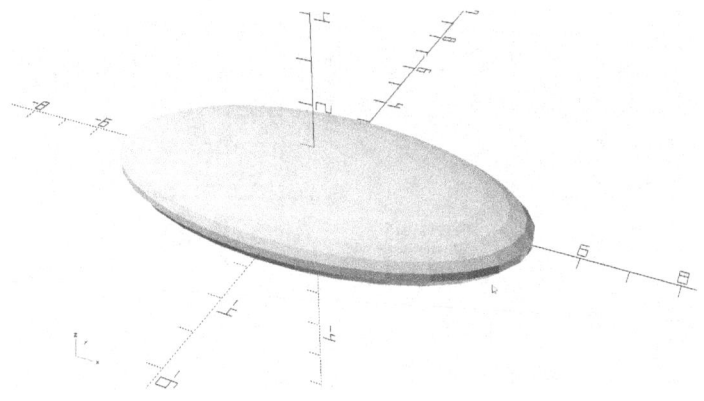

Theoretically, you could create all boxes as a 1-unit cube, and then resize to whatever size box you really want. Here's an example of using this two-step method to create a box with dimensions 10, 5, and 2 units on the X, Y, and Z directions.

```
import openscad as o

# Create and resize a sphere
o.box(1,1,1)
o.resize(10,5,2)

# Send result to OpenSCAD
o.output(o.result())
```

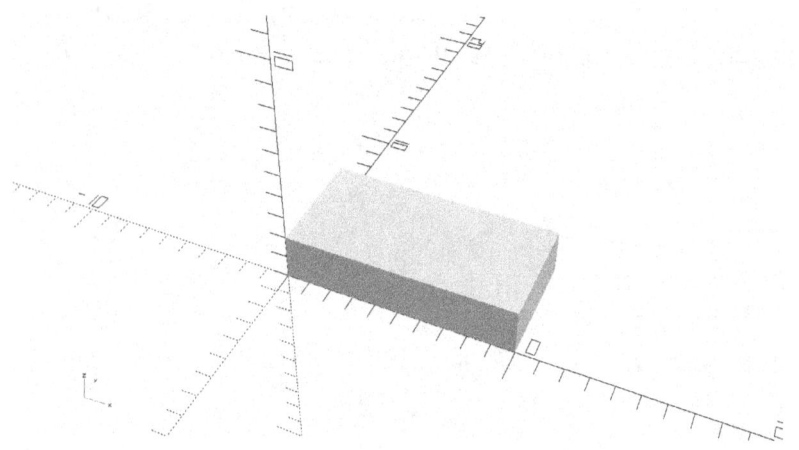

That's all fine and dandy, but where the resize command can really shine is when you might need to stretch a box, or any other construction, in very strange ways. Here's a simple example, where we first rotate a cube and then stretch it to make something like a roofline figure.

```
import openscad as o

# Create, rotate, and resize a box
o.box(1,1,1)
o.rotate(0,45,0)
o.resize(10,10,2)

# Send result to OpenSCAD
o.output(o.result())
```

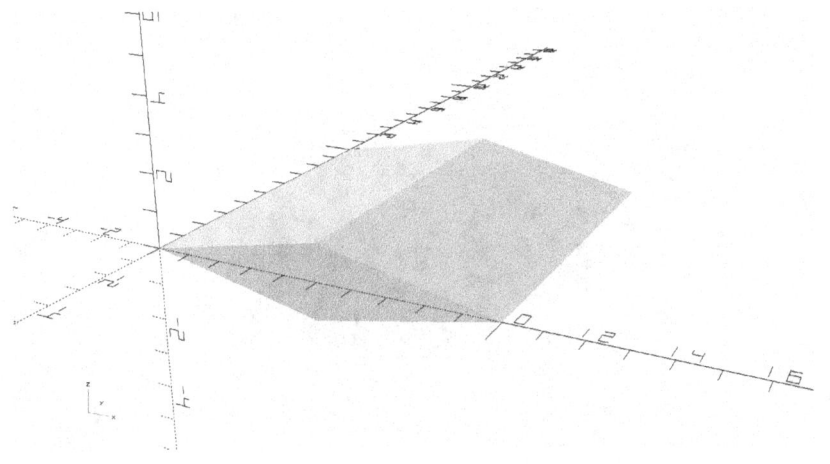

The resize() command is not limited to just box shapes. Any shape in any orientation is fair game for getting stretched in perpendicular directions as desired. Here's one more example, where a cylinder is stretched in all three dimensions.

Python for 3D Printing

```
import openscad as o

# Create, rotate, and resize a cylinder
o.cyl(1,1)
o.resize(10,5,3)

# Send result to OpenSCAD
o.output(o.result())
```

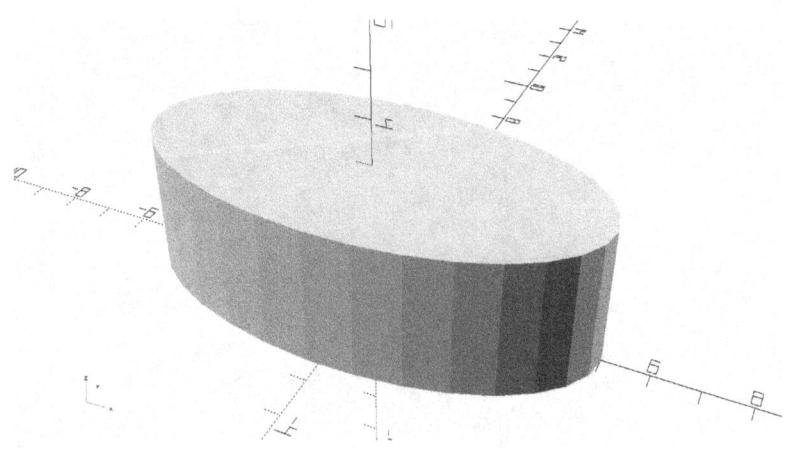

16 - Rotate_extrude

The rotate_extrude command is extremely powerful, a bit mysterious at first, and is easily understood when you break down its action into simple steps. First, a verbal description as clear as I can state it.

Rotate_extrude projects an object onto the X, Y plane, rotates the flat projection 90 degrees around the X axis, and finishes by sweeping the flat geometry around the Z axis by up to 360 degrees to form a volume.

This is one of those commands where a step-by-step example will help clarify how it all works. Let's create a donut or torus using rotate_extrude(). Step one is to create a circle and position it on the X, Y plane. For now, I've commented out the rotate_extrude() command.

Python for 3D Printing

```python
import openscad as o

# Create a "flat" circle
o.cyl(8,o.tiny)

# Position for rotate_extrude
o.translate(10,4,0)

### Flip up and spin
##o.rotate_extrude()

# Send result to OpenSCAD
o.output(o.result())
```

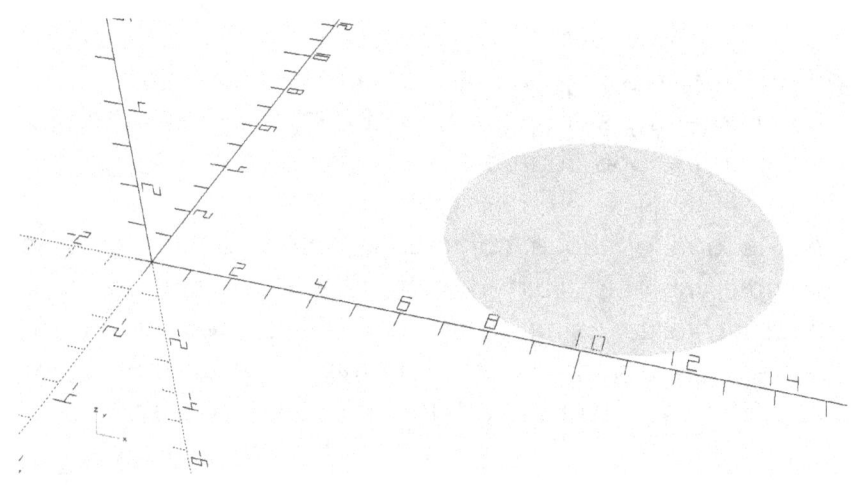

If we had created a sphere with the same diameter and placed it above or below this circle, we'd end up with the same donut result. This is because rotate_extrude takes whatever shape we start with and projects it onto the X, Y plane, as though the Sun were casting its shadow there.

The second step that rotate_extrude takes is to flip our projected shape 90 degrees around the X axis. This kind of stands our circle up on edge, balanced on the X axis line, as shown here.

The final step rotate_extrude takes is to sweep this flat figure around the Z axis to create a volume. In this case we get a nice torus, or donut shape.

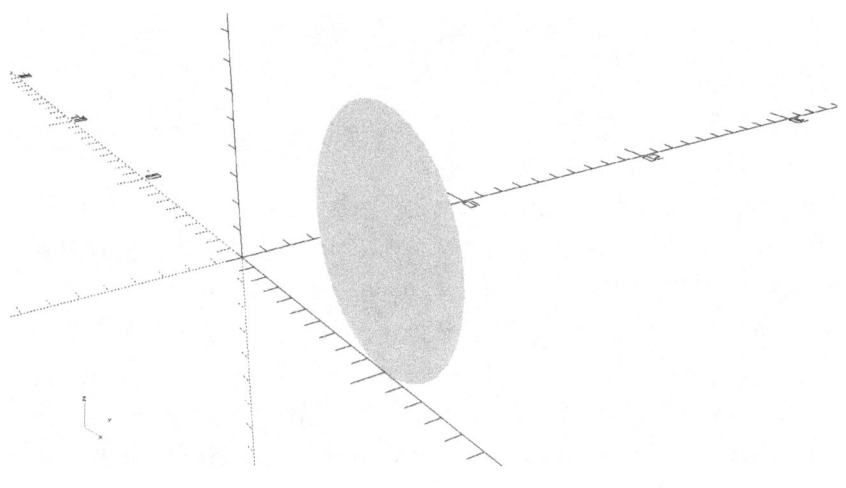

```
import openscad as o

# Create a "flat" circle
o.cyl(8,o.tiny)

# Position for rotate_extrude
o.translate(10,4,0)

# Flip up and spin
o.rotate_extrude()

# Send result to OpenSCAD
o.output(o.result())
```

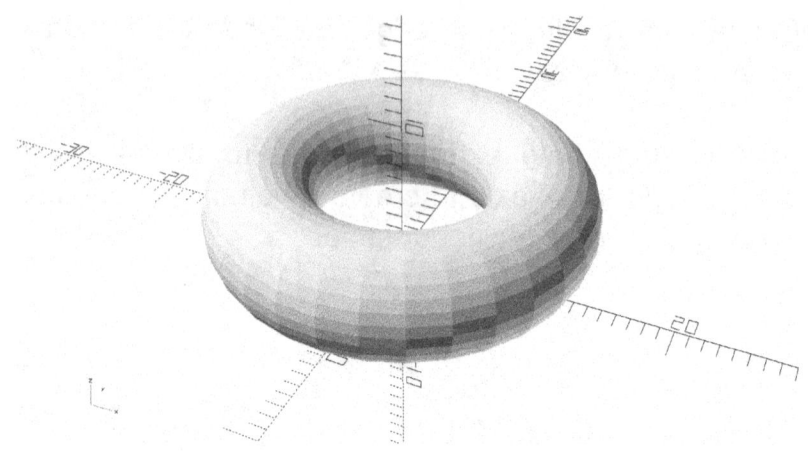

Very interesting shapes can be created by starting with a variety of shapes and then applying the rotate_extrude() command to them. A rectangle can create a tube or washer, a right triangle can create a cone, a rotated circle that projects as an ellipse can form a squashed or stretched donut, and so on. Multiple shapes can be combined into one shape using the union() command (explained in more detail later) and useful, complex, creative shapes can then be formed using rotate_extrude().

Spring shape

The rotate_extrude command is most often used to sweep out a shape in a complete, 360-degree circular volume. But rotate_extrude() does provide an optional parameter for the sweep angle. Setting this parameter to 270 when creating the above donut causes the creation of ¾ of a donut, so it looks like ¼ was bitten out by some hungry coworker.

By using a much smaller sweep angle within a loop that uses the rotate_extrude command repeatedly, we can create a spiraling spring shape, as demonstrated here.

```python
import openscad as o

turns = 5
spacing = 5
segs = []
for i in range(360 * turns):

    # Create a "flat" circle
    o.cyl(3,o.tiny)

    # Position for rotate_extrude
    o.translate(10,spacing * i/360,0)

    # Create 1 degree slice of spiral tube
    o.rotate_extrude(1)

    # Rotate slice to end of spiral tube
    o.rotate(0,0,i)

    # Append the tube slices into a list
    segs.append(o.result())

# Send result to OpenSCAD
o.output(segs)
```

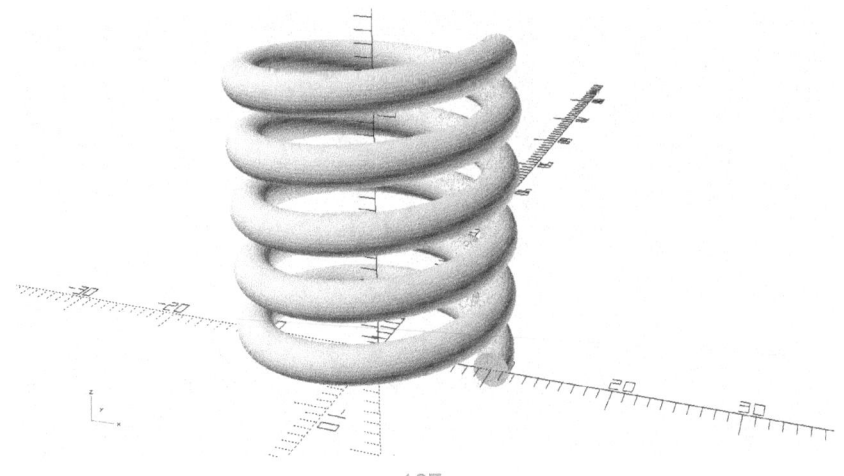

Finding the Holey Grail

I like the power of the rotate_extrude command a lot, so I created an explanatory YouTube video to demonstrate its use in a fun way. It's about three minutes long and can be watched at https://youtu.be/mWbPrCcEeuA

Here's the Python equivalent for the OpenSCAD code I used in that video. The number of fragments is different, just for fun effect, but the rotate_extrude() command is used in the same way.

A polygon is created to outline the edge shape of the grail, and the rotate_extrude() command then picks it up and spins it around to sweep out the grail shape. A couple of spheres are used to punch holes in the side using the difference() command, again just for the nice visual play on words. Sorry, I just can't help it.

```python
import openscad as o

def holy_grail():

    # Set surface smoothness here
    o.fragments = 13

    # Create outline polygon
    thick = 1.5
    points = [
        [0,thick],
        [10,0],
        [10,thick],
        [thick, thick+thick],
        [thick,20],
        [12,30],
        [15,45],
        [15-thick,45],
        [12-thick,31],
        [0,21]]
    o.polygon(points,o.tiny)

    # Rotate up and spin around to make the grail
    o.rotate_extrude()
    regular_grail = o.result()

    # Optionally - make it holey
    o.sphere(10)
    o.translate(5,-7,27)
    hole_1 = o.result()

    o.sphere(14)
    o.translate(-3,-10,35)
    hole_2 = o.result()

    o.difference([regular_grail, hole_1, hole_2])
    holey_grail = o.result()
    return holey_grail

# Call the function
o.output(holy_grail())
```

Python for 3D Printing

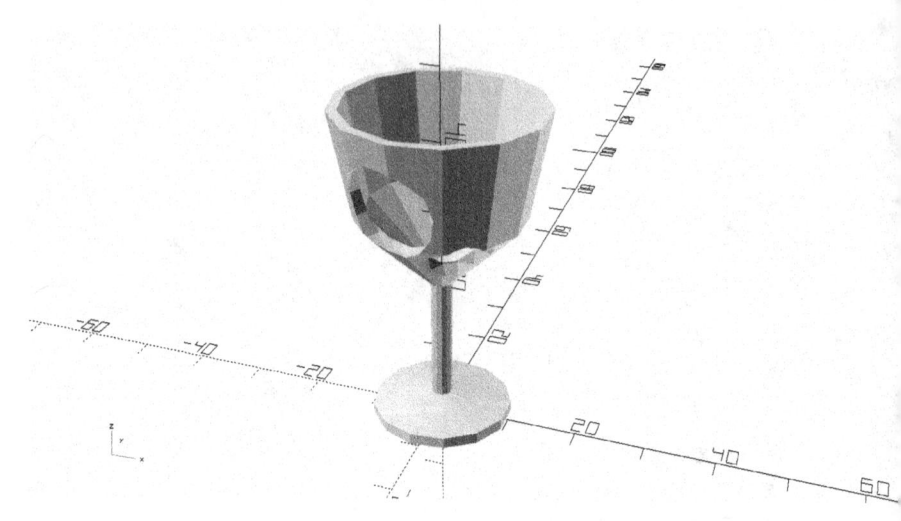

17 - Spiral

The spiral() command takes any object, projects it onto the X,Y plane, then extrudes that flat shape up to a given height, with a given number of turns, and optionally with a scale factor for the top end. Once again, an example helps to clarify how this works. Feel free to play with the various parameters in this example to get a good handle on how the spiral() command works.

For starters, this example creates a twisted square tower, with a quarter turn of rotation, and a height of 10 units. The third parameter is the scale factor. If it's missing, as in this first example, it defaults to 1. This means the square is the same size at the top of the extrusion as at the bottom. A scale factor of 2 would double the size of the top square, and a factor of 0 brings the extrusion to a point.

```
import openscad as o

o.box(10,10,1)
o.translate(-5,-5,0)
o.spiral(.25,10)

# Send result to OpenSCAD
o.output(o.result())
```

Python for 3D Printing

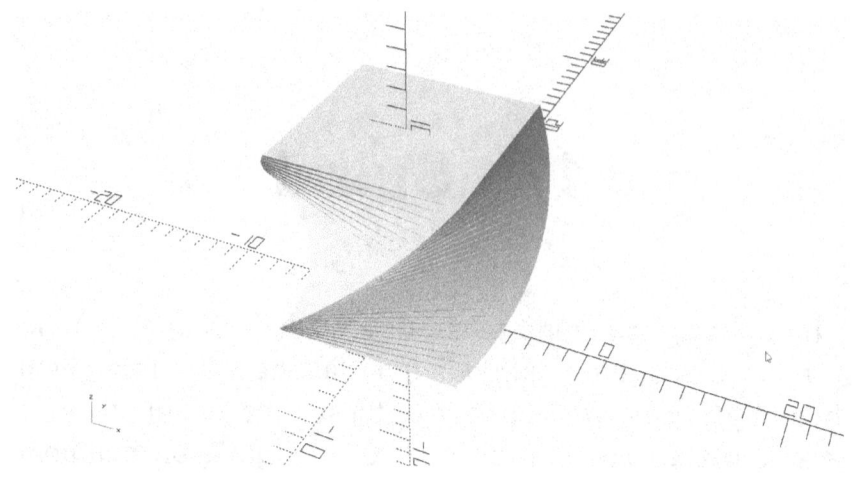

By simply adding the third parameter, scale, to the spiral() command this same figure is changed into a pointed, twisted pyramid. With scale set to 0 the top dimensions are multiplied by zero, and dimensions at each fraction of the way up the tower are linearly proportional to the distance.

```
import openscad as o

o.box(10,10,1)
o.translate(-5,-5,0)
o.spiral(.25,10,0)

# Send result to OpenSCAD
o.output(o.result())
```

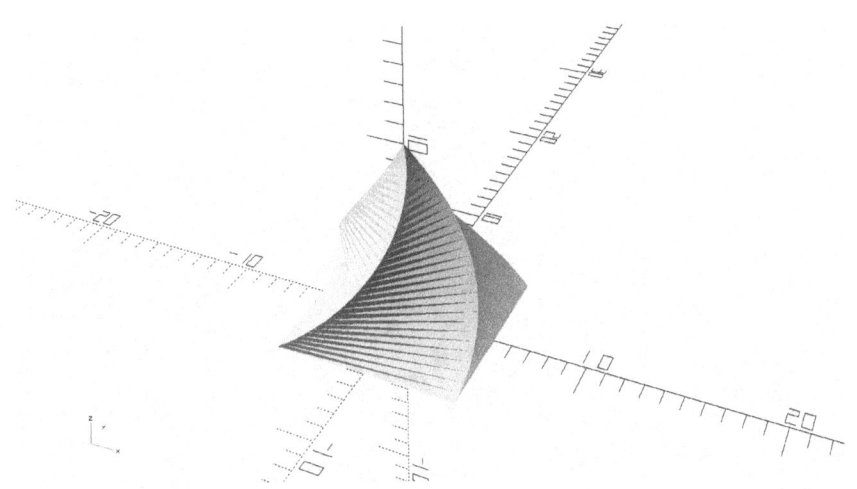

The o.fragments global variable can be used to make the surface smoother if desired. In this next example, to mix things up a bit, I've set o.fragments to make the surface smoother, the scale factor to .5 to create a square at the top that's half as big as at the base, reversed the direction of the spin, and doubled the amount of the spiraling.

```
import openscad as o

o.fragments = 80

o.box(10,10,1)
o.translate(-5,-5,0)
o.spiral(-.5,10,.5)

# Send result to OpenSCAD
o.output(o.result())
```

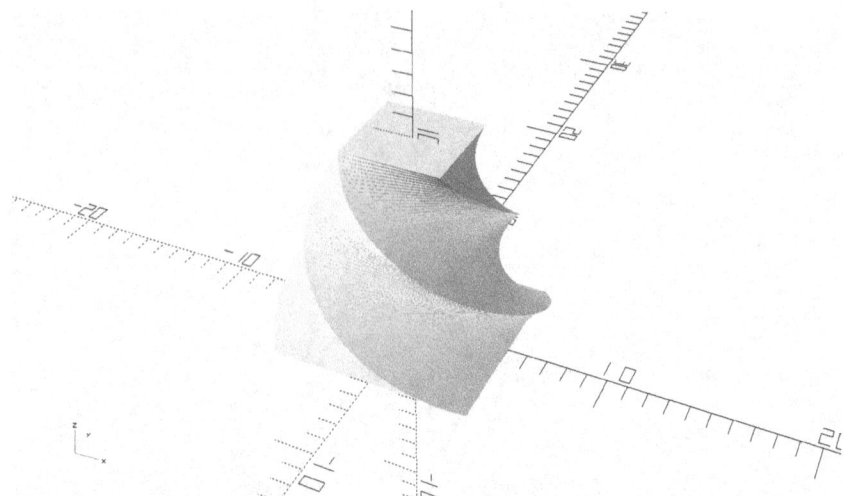

So far we've applied the spiral() command to a square centered on the origin. The spiralling occurs around the Z axis, so an off-center object will create a truly spiralling construction. Here's a type of spiral ramp created by starting with a box moved a little way away from the origin.

```
import openscad as o

o.fragments = 80

o.box(20,1,o.tiny)
o.translate(2,0,0)
o.spiral(1,30)

# Send result to OpenSCAD
o.output(o.result())
```

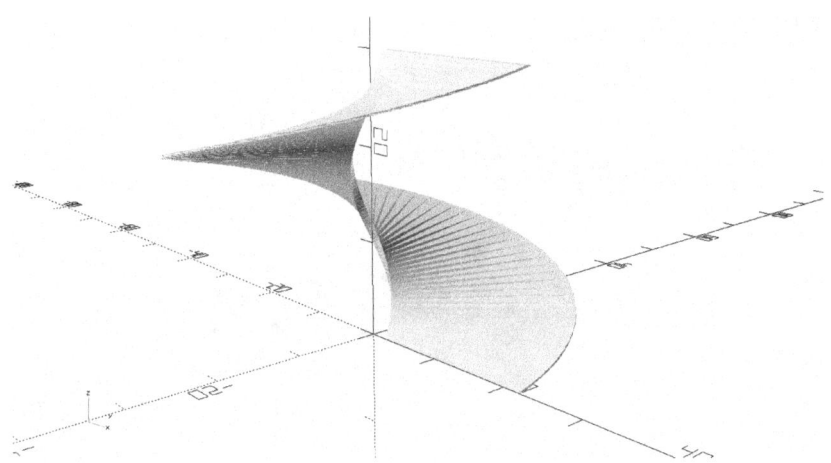

This next example does not use the spiral command, but I include it here to be compared with the spiral ramp. In this case, a series of boxes are constructed, and each is moved into a spiral location that creates a similar construction as the previous ramp. This, however, is more of a spiral staircase.

```python
import openscad as o

for i in range(30):
    o.box(20,5,1)
    o.translate(2,0,0)
    o.rotate(0,0,360 * i / 30)
    o.translate(0,0,i)

# Send result to OpenSCAD
o.output(o.result())
```

Python for 3D Printing

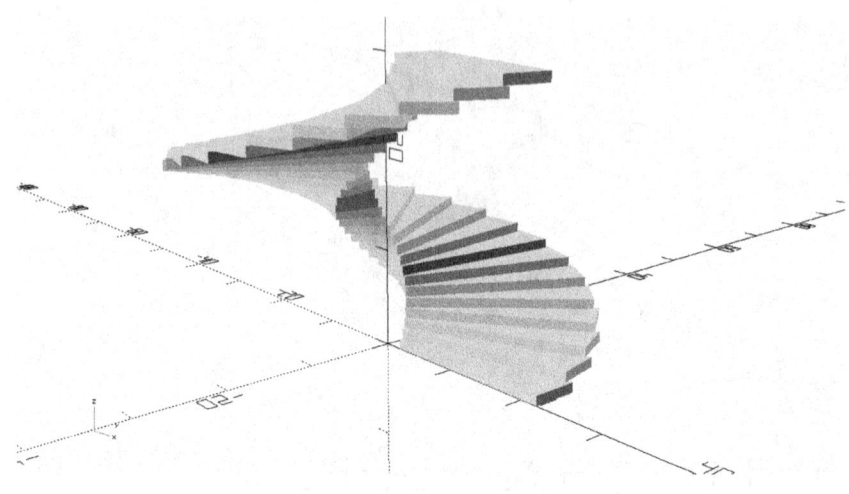

18 - Hull

The hull() command kind of shrink-wraps a list of one or more objects, stretching a new surface between them along the shortest paths possible, wrapping them up into a new shape.

In our first example we'll create two identical cubes, each 5 units long on all edges. One cube remains with its corner at the origin, and the other is translated 10 units along the Y axis. The hull() command wraps these two cubes to create a new rectangular solid that is longer in the Y direction.

```python
import openscad as o

# Create two indentical cubes
o.box(5,5,5)
o.translate(0,0,0)

o.box(5,5,5)
o.translate(0,10,0)

# Wrap them with hull
o.hull(o.result())

# Send result to OpenSCAD
o.output(o.result())
```

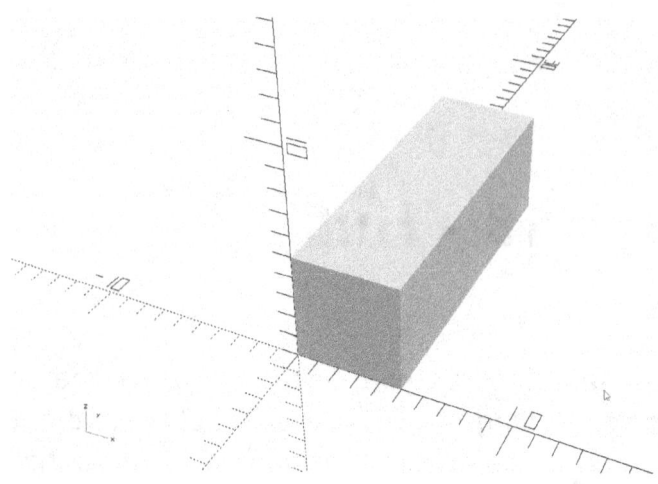

This next example similarly creates two cubes, but one is translated diagonally out along the X, Y plane. The hull() wraps the two into a pointy crystal-like shape. Note that the two cubes are saved in their own variables and the hull() command combines them in a list. Compare this to the way the first two cubes were more directly handed to the hull() command in the first example. Either way works fine.

```python
import openscad as o

# Create two cubes
o.box(5,5,5)
o.translate(0,0,0)
cube1 = o.result()

o.box(5,5,5)
o.translate(10,10,0)
cube2 = o.result()

# Wrap them with hull
o.hull([cube1,cube2])

# Send result to OpenSCAD
o.output(o.result())
```

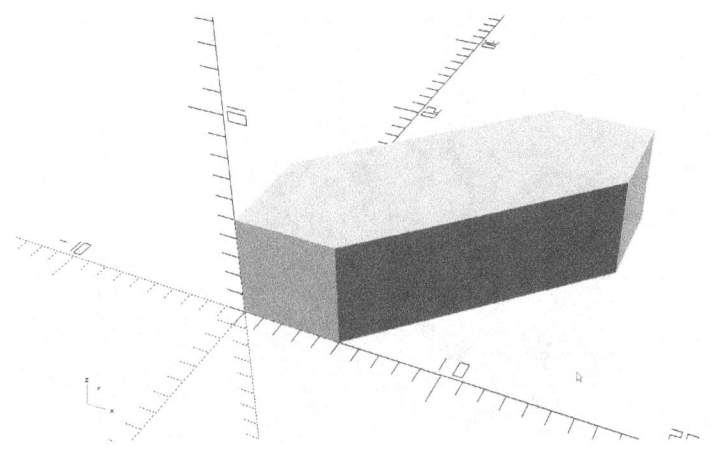

We're still not quite done with those two cubes. In this next example the second cube is translated up, over, and out in all three directions. As a result, the hull() command shows the cube faces and points more clearly as it wraps them at this skewed angle.

```
import openscad as o

# Create two cubes
o.box(5,5,5)
o.translate(0,0,0)

o.box(5,5,5)
o.translate(10,10,10)

# Wrap them with hull
o.hull(o.result())

# Send result to OpenSCAD
o.output(o.result())
```

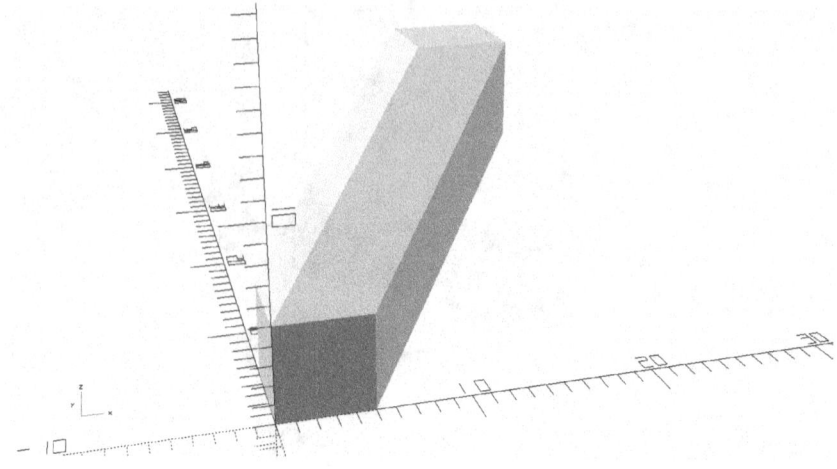

The hull() command is, of course, not limited to shrink wrapping cubes. Here's an example where two spheres are created, one large and one small, and the hull() command wraps them up into a hot-air balloon shape.

```
import openscad as o

# Create larger sphere
o.sphere(20)
s1 = o.result()

# Create smaller sphere
o.sphere(3)
o.translate(0,0,-20)
s2 = o.result()

# Hull around the pair of spheres
o.hull([s1,s2])

# Send result to OpenSCAD
o.output(o.result())
```

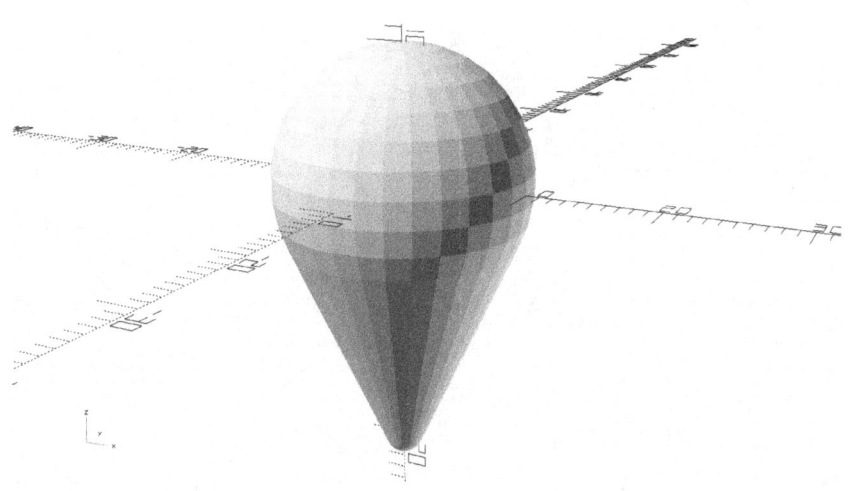

Hull() wrapping spheres and cylinders provides a technique for creating rounded corners and edges on box shapes. To see how this works, let's first create 8 spheres and move them out in space to the corners of a cubical volume.

```python
import openscad as o

# Create 8 corner spheres
for x in range(-10,20,20):
    for y in range(-10,20,20):
        for z in range(-10,20,20):
            o.sphere(10)
            o.translate(x,y,z)

# Wrap everything in a hull
#o.hull(o.result())

# Send result to OpenSCAD
o.output(o.result())
```

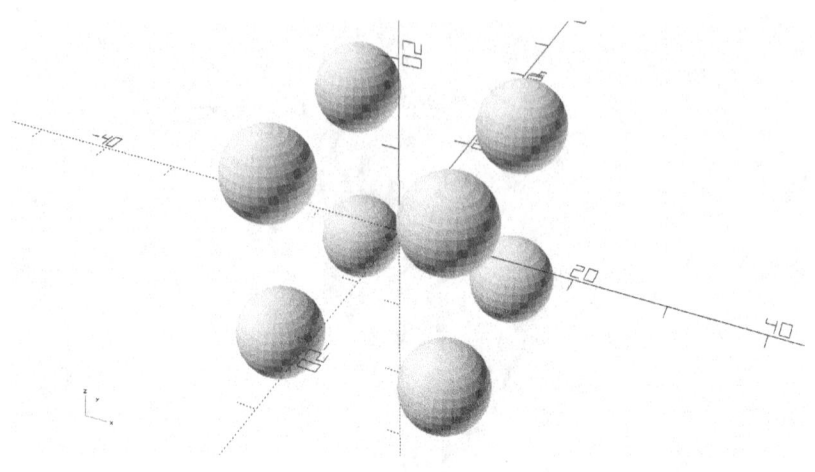

Now let's uncomment the hull() command to shrink wrap all 8 balls into a round-edged and round-cornered box shape.

```
import openscad as o

# Create 8 corner spheres
for x in range(-10,20,20):
    for y in range(-10,20,20):
        for z in range(-10,20,20):
            o.sphere(10)
            o.translate(x,y,z)

# Wrap everything in a hull
o.hull(o.result())

# Send result to OpenSCAD
o.output(o.result())
```

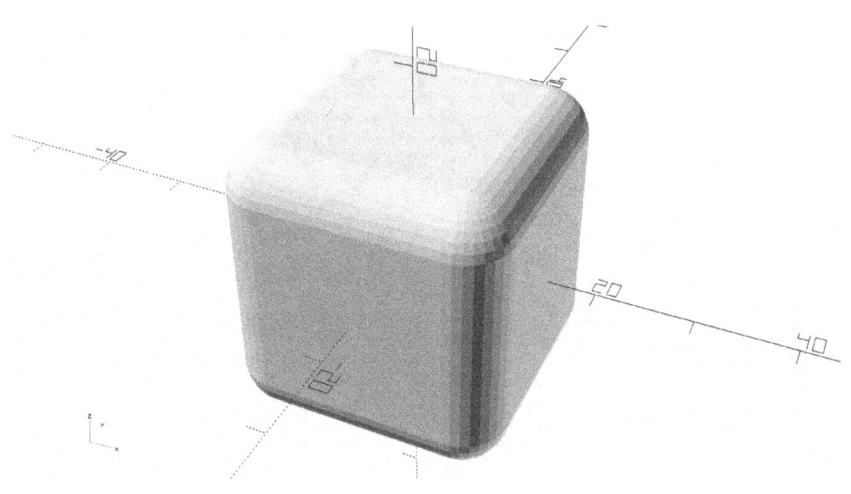

Be aware that you can use hull() on cylinders to create rounded side edges while leaving the top and bottom flat. There are lots of other creative ways hull() can simplify some complex design challenges.

19 - Minkowski

The minkowski() function, I presume, was named after someone named Minkowski. The OpenSCAD documentation doesn't go into detail explaining how it works or where it came from, but with a little experimentation it all starts to make sense. The minkowski() command turns out to be very powerful indeed.

Start with the following code, where several objects have been commented out. As it stands, this code just creates a simple box, as shown.

```python
# minkowski.py
import openscad as o

# Uncomment any two of these at a time...
o.box(10,20,3)
##o.sphere(5)
##o.cyl(5,5)
##o.box(10,10,10)

# Grab the combination of objects
multiple_objects = o.result()

# Process together using minkowski()
o.minkowski(multiple_objects)

# Send results to OpenSCAD
o.output(o.result())
```

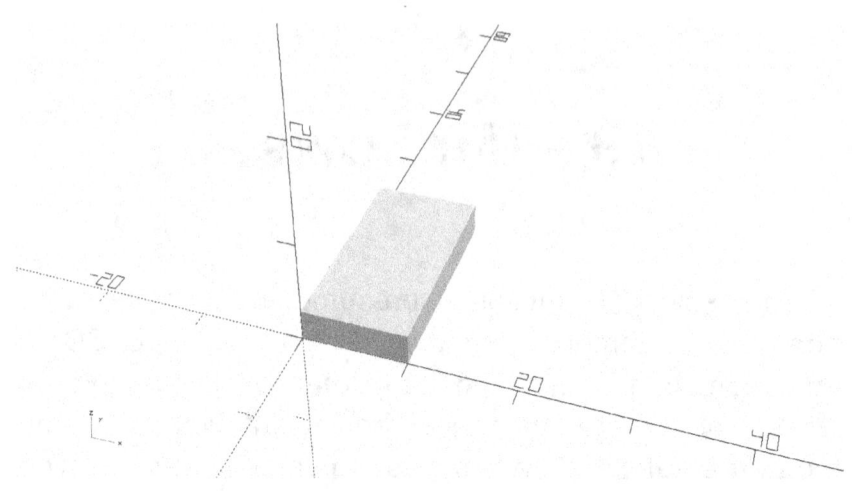

When more than one object is combined together using minkowski(), each object puffs the results out along each of the three axes by its dimension in that direction. That's the easiest way I've figured out how to describe the action. This next example helps pin it down. Uncomment the sphere() command to see how the original box has sides, rounded edges, and rounded corners resulting from the sphere "puffing out" in each direction based on the radius of the sphere.

```python
# minkowski.py
import openscad as o

# Uncomment any two of these at a time...
o.box(10,20,3)
o.sphere(5)
##o.cyl(5,5)
##o.box(5,5,5)

# Grab the combination of objects
multiple_objects = o.result()

# Process together using minkowski()
o.minkowski(multiple_objects)

# Send results to OpenSCAD
o.output(o.result())
```

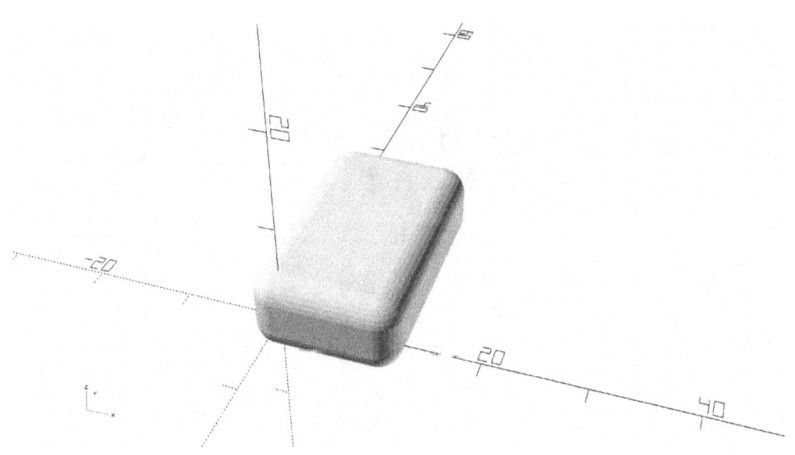

Similar to the hull() command, this is another very handy way to create rounded edges and corners on shapes. If you want the box to stay the same size from face to face as originally sized, subtract the radius of the sphere from each of its starting dimensions.

As a next experiment, enable the cylinder shape instead of the sphere. This results in rounded vertical edges and flat top and bottom surfaces to be added to the box. If you think about it, the cylinder has the same type of geometry, where the vertical "edges" are round, and the top and bottom surfaces are flat.

```
# minkowski.py
import openscad as o

# Uncomment any two of these at a time...
o.box(10,20,3)
##o.sphere(5)
o.cyl(5,5)
##o.box(5,5,5)

# Grab the combination of objects
multiple_objects = o.result()

# Process together using minkowski()
o.minkowski(multiple_objects)

# Send results to OpenSCAD
o.output(o.result())
```

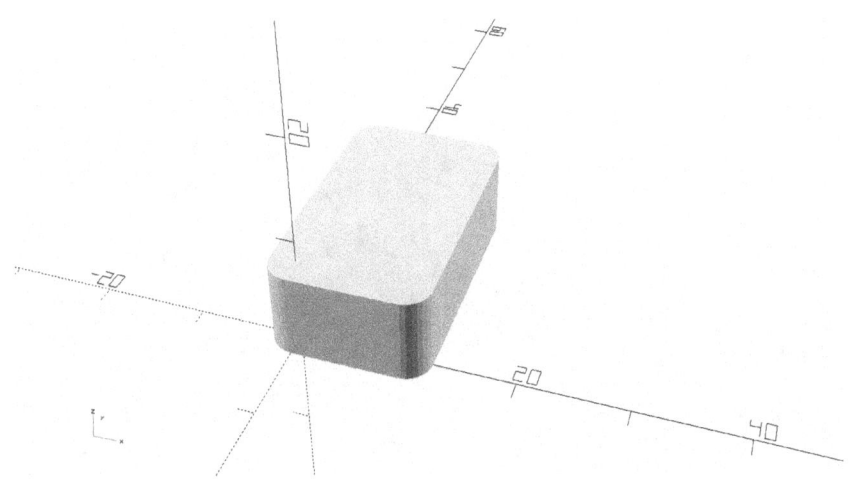

It's informative to combine two boxes using minkowski(), although this is not an often used way to create boxes. The following code enables two boxes to add to each other using minkowski(), resulting in a bigger box that is the sum of the first two in each dimension.

```python
# minkowski.py
import openscad as o

# Uncomment any two of these at a time...
o.box(10,20,3)
##o.sphere(5)
##o.cyl(5,5)
o.box(5,5,5)

# Grab the combination of objects
multiple_objects = o.result()

# Process together using minkowski()
o.minkowski(multiple_objects)

# Send results to OpenSCAD
o.output(o.result())
```

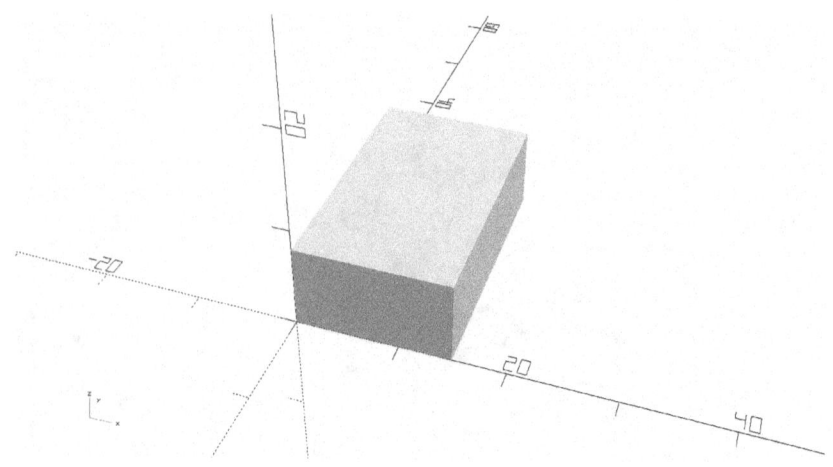

More than two objects can be combined using the minkowski() command. All the shapes in the passed list are combined in the various directions to expand the original shape. For this last example, all 4 objects are enabled.

Also, to show that the minkowski() command can accept either one result or a list of objects, each of the 4 shapes is stored in a variable, and they are all handed to the minkowski() command in a list. This provides a great way to experiment further with various combinations of the objects. Just edit the list to include them as desired.

```python
# minkowski.py
import openscad as o

# Create four named objects
o.box(10,20,3)
box1 = o.result()

o.sphere(5)
sphere = o.result()

o.cyl(5,5)
cylinder = o.result()

o.box(5,5,5)
box2 = o.result()

# Combine two or more in a list
the_list = [box1,sphere,cylinder,box2]

# Process together using minkowski()
o.minkowski(the_list)

# Send results to OpenSCAD
o.output(o.result())
```

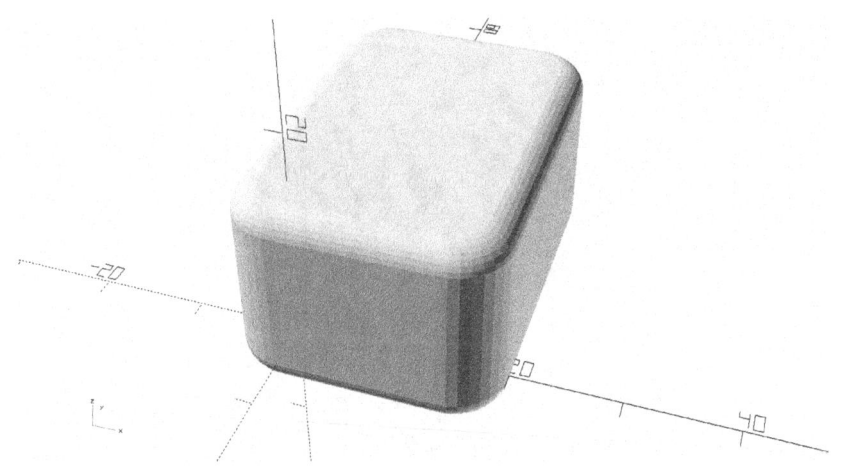

Python for 3D Printing

20 - Mirror

The mirror() command reverses all geometry defined by a plane that passes through the origin. A common way to use mirror() is to create an object to either side of the Y,Z plane, and then to create a mirror image reflection of the object on the other side of the Y,Z plane.

The mirror() command expects three parameters, representing a vector that defines the plane for the mirror action. So, as described above, mirror(1,0,0) designates a vector that points along the X axis, and this completely defines the plane that passes through the origin and coincides with the Y,Z plane. Note that mirror(100,0,0), also defines the same plane, as it's the direction of the vector, and not its magnitude, that is important.

You can reflect through any plane that passes through the origin. As an example, mirror(33,44,55) defines a plane through the origin that is perpendicular to this vector. However, good luck with trying to visualize the result of using this command! It can be done, but reflecting in a simpler plane is easier.

Here's an example that creates text at some floating angle in space to the right of the Y, Z plane, and then also creates its mirror image on the other side of the Y, Z plane.

```python
import openscad as o

# Red text floating in space
o.text("Python for OpenSCAD",height=5)
o.translate(15,10,0)
o.rotate(0,-22,22)
o.color("red")

# Light-blue text mirrored in Y-Z plane
o.text("Python for OpenSCAD",height=5)
o.translate(15,10,0)
o.rotate(0,-22,22)
o.color("lightblue")
o.mirror(1,0,0)

# Send result to OpenSCAD
o.output(o.result())
```

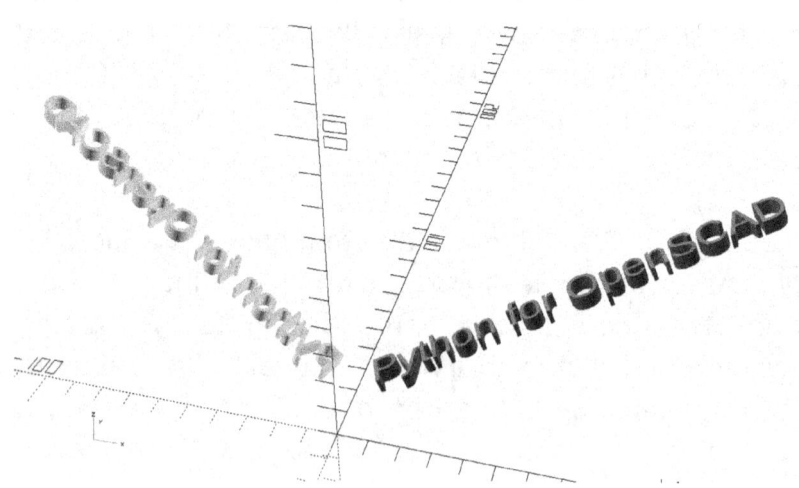

Left-handed threads using a library

One useful application of the mirror() command is to create left-handed threads on a bolt. Of course, this begs the question of how to create right-handed threads in the first place. There are OpenSCAD libraries available for many purposes, and creating threads is most easily accomplished by using one of these libraries. This example shows how to use a library, and then how to mirror the results to create a left-handed bolt. (If you want a normal, right-handed bolt, simply leave out the mirror() command.)

Two new commands are used in this example to help interface with the OpenSCAD library. They provide some required flexibility for special circumstances. The startup() command adds its string parameter to the translated OpenSCAD code to allow correct application of OpenSCAD's "use" statement. This is how to include a given library file. Be sure to place a copy of the threads.scad library file in the same folder as your project. You can search for this library on the Internet.

A very flexible command is literally called literally(), used here to insert a very specifically designed string command in our OpenSCAD output, in this case to tell the library that we want a metric thread with specific diameter, pitch, and length parameters. Other libraries might require other unique strings or command information, and the literally() command lets you insert whatever you might need. Note that triple quotes are used, allowing line feeds, quotes, and pretty much any text or characters required to be inserted into the OpenSCAD output. You'll need to study and understand the unique requirements of any library code to know exactly how to use literally(), but there you go.

In this example the library is used to create a spiral of threads, and other commands create the bolt and head parts. They are combined to create a complete, right-handed bolt centered at the origin. To create the mirror image, the bolt is translated away from the origin, mirrored through the Y, Z plane, and then translated back to its original position at the origin. The right-handed version of this bolt is also shown, and it was created by commenting out the mirror() command.

```python
import openscad as o
o.fragments = 64

o.startup("use <threads.scad>;")

# Threaded part
o.literally("""
metric_thread (diameter=8, pitch=1, length=12);
""")
threads = o.result()

# Unthreaded part
o.cyl(8,12)
o.translate(0,0,-12)
bolt = o.result()

# Head
o.regular_polygon(6,8,4)
o.translate(0,0,-16)
head = o.result()

# Combine the parts
o.union([threads, bolt, head])
o.color("silver")

# Move up and away from origin
o.translate(10,0,16)

# Mirror in Y-Z plane
o.mirror(1,0,0)

# Move back to origin
o.translate(10,0,0)

# Send result to OpenSCAD
o.output(o.result())
```

Python for 3D Printing

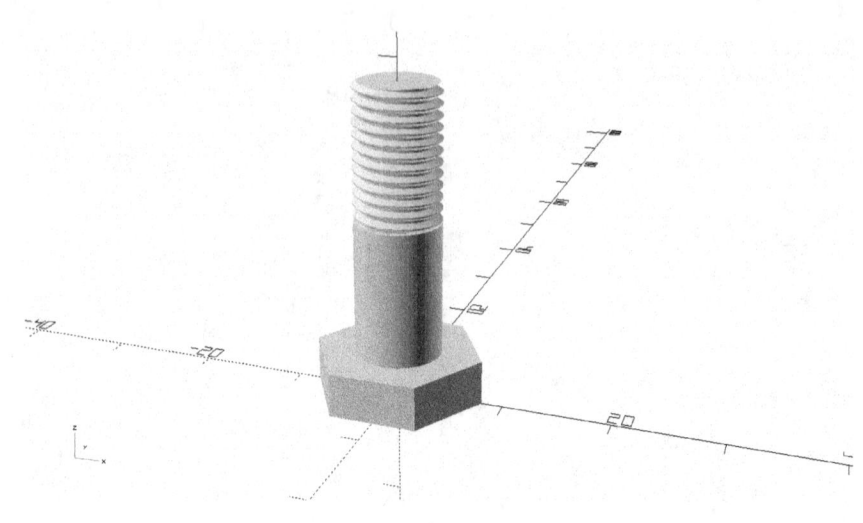

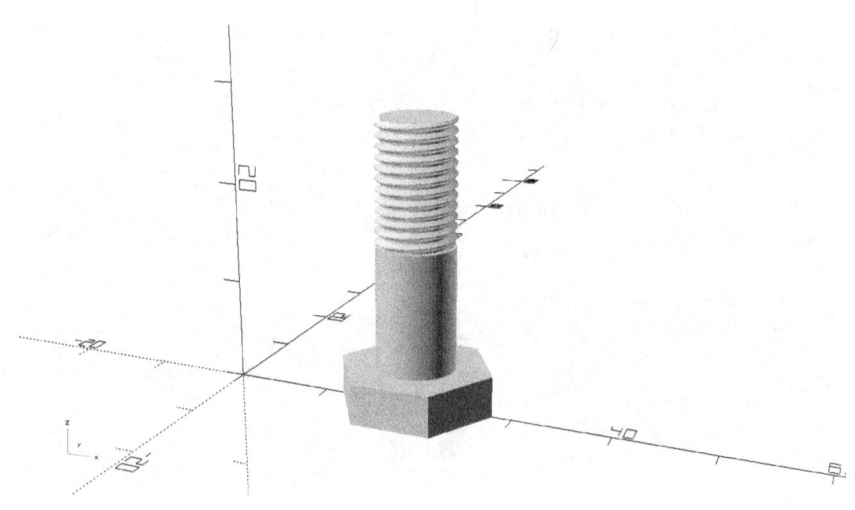

21 - Projection

The projection() command takes any shaped object and projects it onto the X,Y plane. I like to think of this as though the sun were shining directly down and casting a shadow of the object. Or, if part or all of the object is below the X, Y plane, the sun is shining up from below.

There is one parameter to pass, and that's how thick the projection should be extruded in the Z axis direction, once the object is flattened onto the X, Y plane. For a very, very flat shape, use o.tiny for this parameter.

This first example creates a flat circle, rotates it in space, then lifts it up along the Z axis. A second flat circle is created identically, except that it is then projected onto the X, Y plane. It ends up as an ellipse because of the rotations involved.

Python for 3D Printing

```
import openscad as o

# Create a "flat" circle tilted in space
o.cyl(10,o.tiny)
o.rotate(45,0,-45)
o.translate(0,0,15)

# Also create projection of same shape
o.cyl(10,o.tiny)
o.rotate(45,0,-45)
o.translate(0,0,15)
o.projection(o.tiny)
o.color("red")

# Send result to OpenSCAD
o.output(o.result())
```

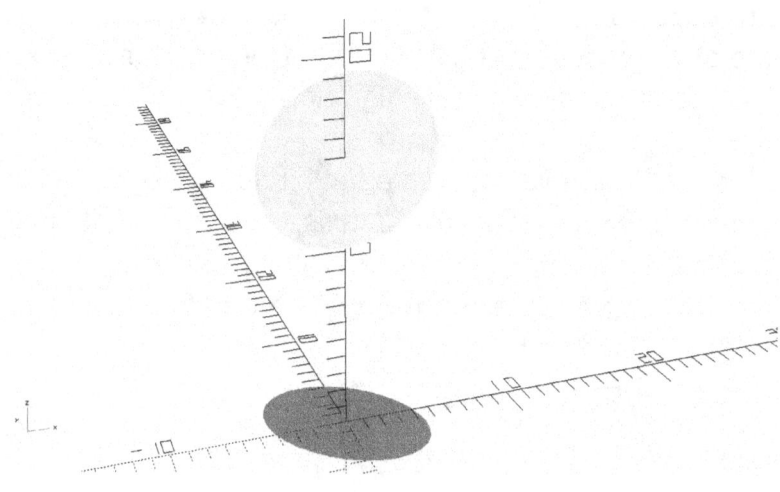

To demonstrate adding thickness using the height parameter, the following code is identical except for changing o.tiny to 5 units.

```
import openscad as o

# Create a "flat" circle tilted in space
o.cyl(10,o.tiny)
o.rotate(45,0,-45)
o.translate(0,0,15)

# Also create projection of same shape
o.cyl(10,o.tiny)
o.rotate(45,0,-45)
o.translate(0,0,15)
o.projection(5)
o.color("red")

# Send result to OpenSCAD
o.output(o.result())
```

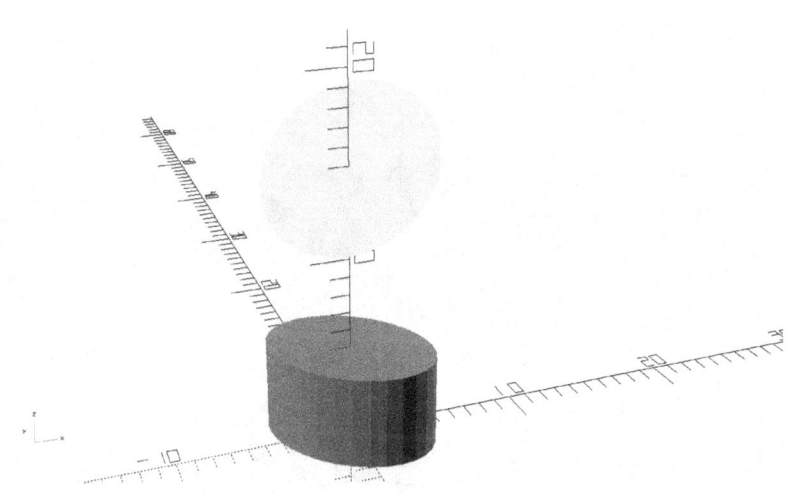

Python for 3D Printing

This next example gives you a hint at all the possibilities when using projection() on other shapes at various orientations in space. A cylinder with diameter 10 and height 10 is floated above the origin, and a second copy of the same shape cylinder is projected onto the X, Y plane, resulting in a tablet shape.

```
import openscad as o

# Create a cylinder tilted in space
o.cyl(10,10)
o.rotate(45,0,-45)
o.translate(0,0,15)

# Also create projection of same shape
o.cyl(10,10)
o.rotate(45,0,-45)
o.translate(0,0,15)
o.projection(3)
o.color("red")

# Send result to OpenSCAD
o.output(o.result())
```

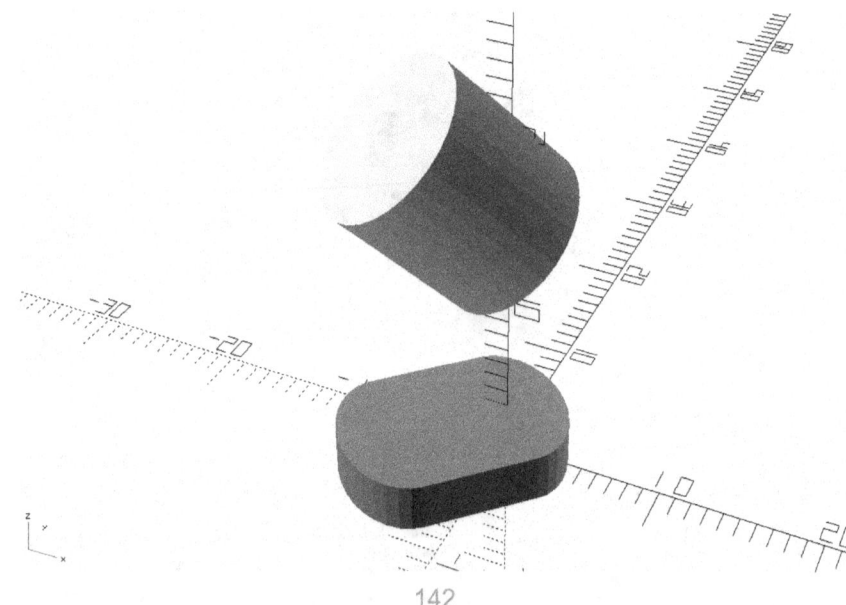

As a side thought, it's possible the creators of OpenSCAD decided to add several commands to create true 2D shapes, with a false height of 1 unit (somewhat confusingly), in order to avoid problems with 3D objects projecting imperfect shapes. For example, a circle with any thickness at all theoretically casts a projection like our tablet shape above, when tilted, rather than to project a perfect ellipse as desired. I decided to have all commands in openscad.py create true 3D objects. The o.tiny thickness of cylinders and other shapes completely eliminates this distortion problem in a much easier to understand way - o.tiny is so small that no measurable effect of this thickness will show up in any normal projections.

Cube into a hexagon

If you're careful, plan ahead, and know what you're doing, you can create some very interesting geometrical figures from simpler constructions. As one example, here's a cube rotated in space exactly the right amount to place one corner directly above its opposite corner. The projection of this cube results in a perfect hexagon.

```
import openscad as o
import math

rotx = 45
roty = math.atan(1/math.sqrt(2))*180/math.pi
rotz = 0

# Create a cube tilted in space
o.box(10,10,10)
o.rotate(roty,rotx,rotz)
o.translate(0,0,15)

# Also create projection of same shape
o.box(10,10,10)
o.rotate(rotx,roty,rotz)
o.translate(0,0,15)
o.projection(o.tiny)
o.color("red")

# Send result to OpenSCAD
o.output(o.result())
```

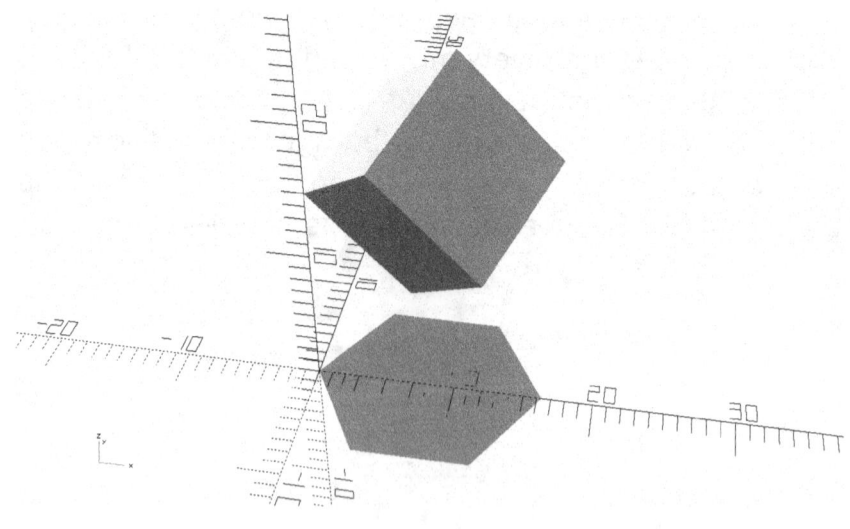

22 - Slice

The slice() command is exactly the same as the projection() command, only different. Well, they are similar in some ways.

slice() comes in handy when a cross section of any shape is desired. Imagine creating a 3D bust of a famous figure, then slicing the model to create thin cross-section slabs from bottom to top, cutting those shapes out of cardboard, wood, or plastic, and gluing them together to make the bust as an art form. The slice() command is perfect for this, and many other tasks.

The slice() command works only through the X,Y plane, and its one parameter adds a height, or vertical thickness to the result. As usual, if you want a very thin slice, use o.tiny for the height, but if you want slices representing something like cardboard slabs, use a small height value, probably the same as the increment used to position the object vertically for each slice.

Let's start with a very simple case, to make sure you understand how slice() works. A reference sphere is created and made nearly transparent. An identical sphere is then sliced, using o.tiny to create just a flat slice.

```
import openscad as o

# Create a sphere
o.sphere(10)
o.color("gold",.3)

# Slice through same shape
o.sphere(10)
o.slice(o.tiny)
o.color("blue")

# Send result to OpenSCAD
o.output(o.result())
```

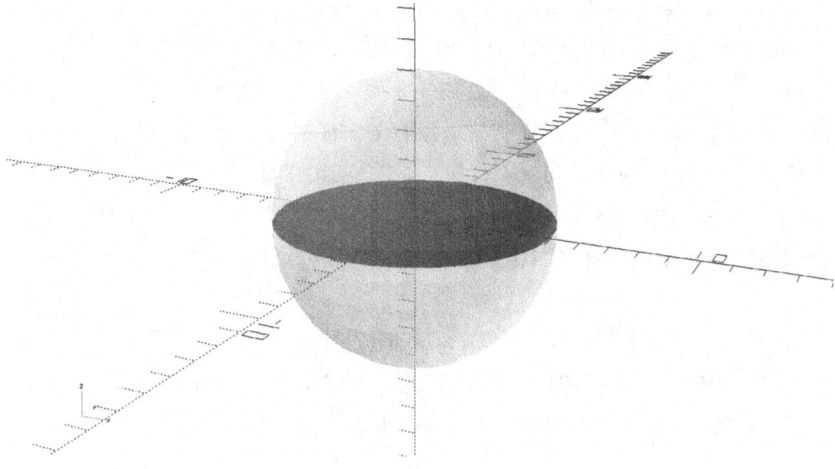

If we change the slice() command's height parameter from o.tiny to 5, we end up with a dramatically different cylinder result.

```python
import openscad as o

# Create a sphere
o.sphere(10)
o.color("gold",.3)

# Slice through same shape
o.sphere(10)
o.slice(5)
o.color("blue")

# Send result to OpenSCAD
o.output(o.result())
```

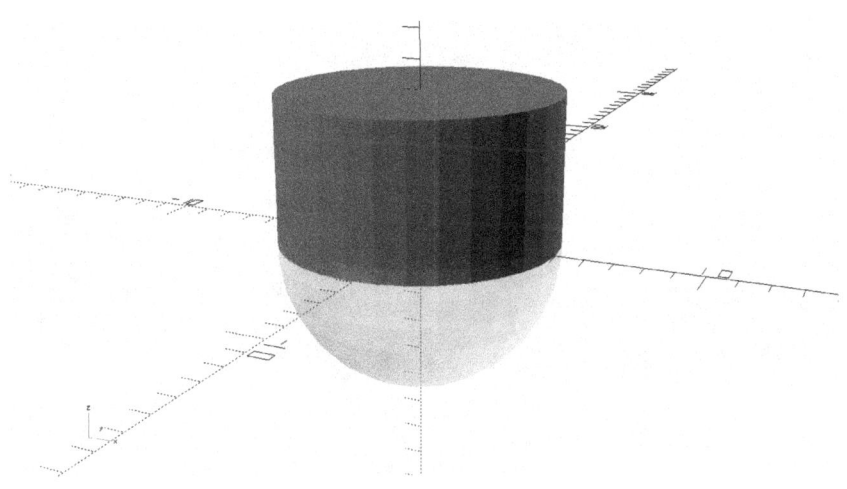

The next example creates a box shape oriented in space at a funny angle partly above and partly below the X, Y plane. A second identical box in the same orientation is then sliced. If the view is rotated to look straight down on the slice, you'll see the result is a skewed rectangle, as though stretched by pulling on opposite corners.

Python for 3D Printing

```
import openscad as o

# Create a box cutting through X,Y plane
o.box(5,10,20)
o.translate(-2,-4,-6)
o.rotate(22,33,44)
o.translate(10,10,0)
o.color("gold",.3)

# Slice through same shape
o.box(5,10,20)
o.translate(-2,-4,-6)
o.rotate(22,33,44)
o.translate(10,10,0)
o.slice(o.tiny)

# Send result to OpenSCAD
o.output(o.result())
```

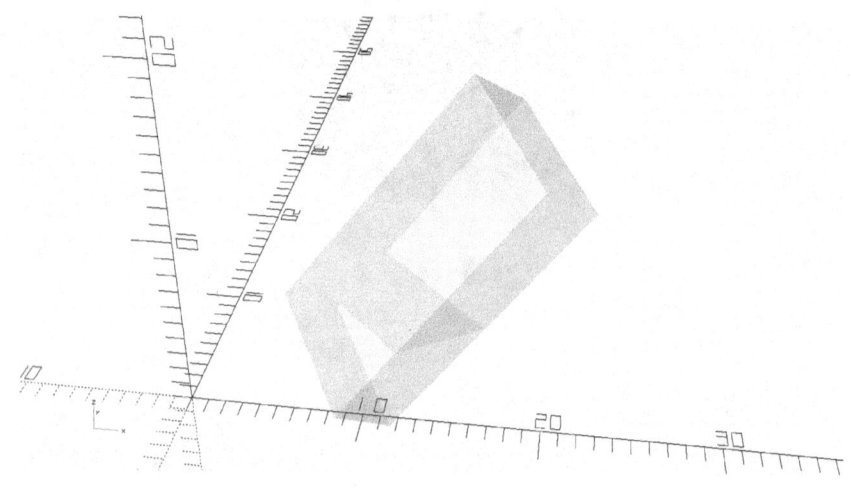

This final demonstration of the slice() command creates a pyramid and slices it into thin slabs. The $t parameter was added to allow animation. The first translate() command shifts the pyramid such that the slice location is positioned in the X,Y plane, and the second translate() command shifts the whole thing back to its starting position.

The pyramid construction is isolated in a function to avoid repetition. The pyramid is first created in a mostly transparent form, so you can see where the slicing action is happening, and the second call to pyramid allows the actual slice() processing.

```python
import openscad as o

# Create a pyramid
def pyramid():
    o.box(10,10,1)
    o.translate(-5,-5,0)
    o.spiral(0,10,0)

pyramid()
o.color("cyan",.2)

# Try animating at FPS: 30 Steps: 100
pyramid()
o.translate(0,0,"-$t*10")
o.slice(.3)
o.translate(0,0,"$t*10")

# Send result to OpenSCAD
o.output(o.result())
```

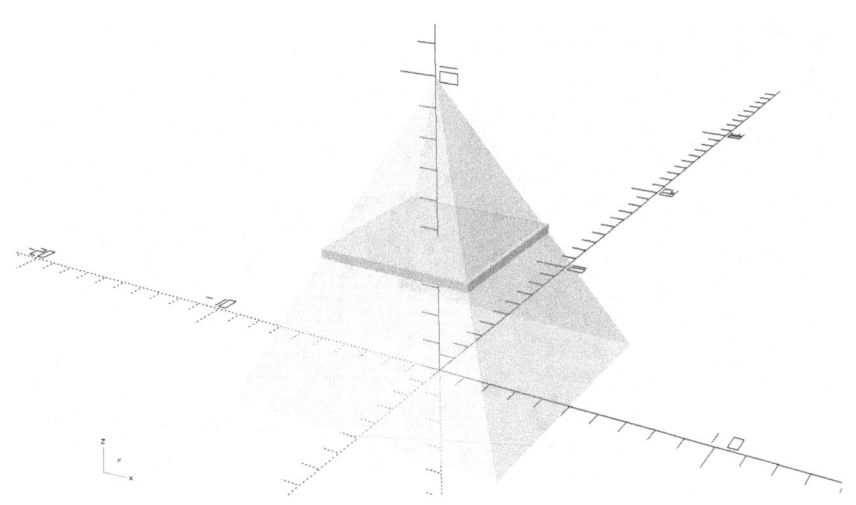

To see the animation in action, from OpenSCAD's View menu select Animate. In the "FPS:" field at the bottom of the display enter 30, and in the "Steps:" field enter 100. You might want to change these values, along with the height parameter in the slice() command, to create an animation to your liking.

23 - Offsets

The offset group of commands creates externally or internally adjusted outlines of the projection of a shape onto the X, Y plane. The outlines can have straight, rounded, or chamfered corners, based on the version of the offset command used.

The first parameter is the distance outward or inward to adjust the outline, with positive values moving outward, and negative values moving inward. The second parameter is height, which adds a thickness to make our offset a 3D object.

Take a close look at this code to better understand how these commands work. As shown in this version of the code, the offset_round() command with a positive adjustment distance is enabled, and the other five command combinations are commented out. The first screen grab shows the polygon shape before any of these offsets are applied, and the rest of the screen grabs show the various offsets in action, one at a time.

```python
import openscad as o

# Keep curved surfaces smooth
o.fragments = 100

# Create a polygon outline
p1 = [0,0]
p2 = [10,0]
p3 = [10,10]
p4 = [5,5]
p5 = [0,15]
outline = [p1,p2,p3,p4,p5]

# Polygon with no offset
o.polygon(outline,3)
base = o.result()

# Same polygon outline, ready for offset
o.polygon(outline,o.tiny)

# Add one of these external offsets
o.offset_round(2,2)
##o.offset_straight(2,2)
##o.offset_chamfer(2,2)

# (Or) add one of these internal offsets
##o.offset_round(-1,4)
##o.offset_straight(-1,4)
##o.offset_chamfer(-1,4)

# Color the offset pink
o.color("pink")
offset = o.result()

# Send combined results to OpenSCAD
o.output([base,offset])
```

Here's the polygon shape with none of the offset commands enabled.

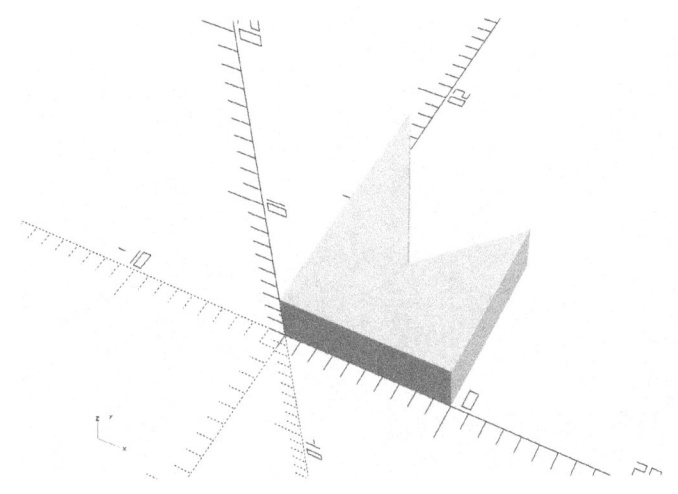

Here's a combination of the original polygon plus he shape created when offset_round(2,2) is applied. Note that the original polygon has a height of 3 units, and the offset outline has a height of 2 units, so they can both be seen together at the same time. The color of the offset has been changed to pink to enhance the visual difference.

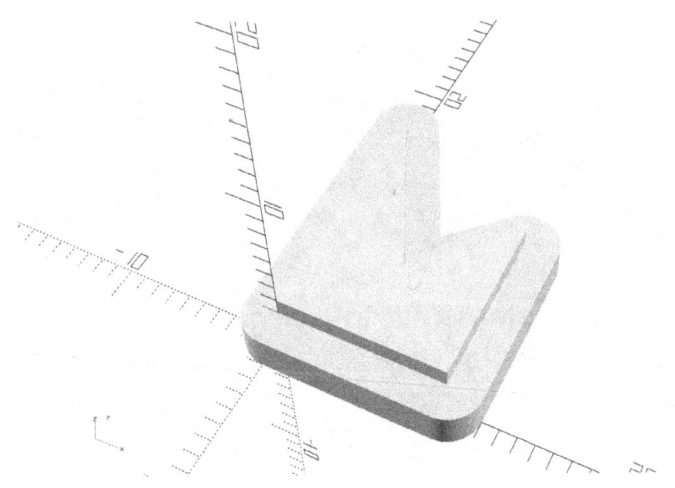

Next is the result of using offset_straight(2,2).

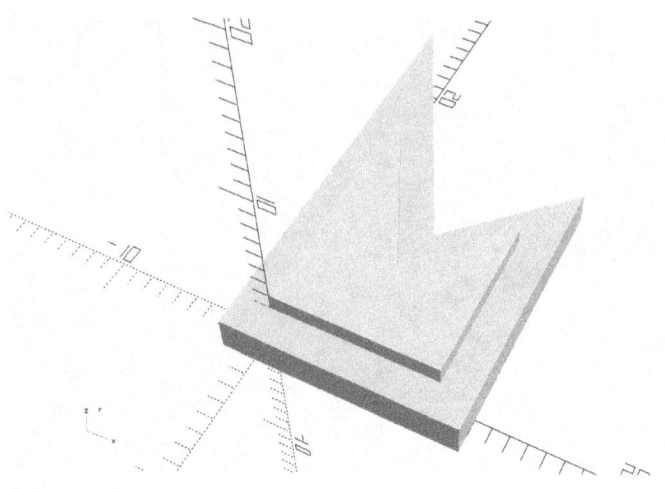

offset_chamfer(2,2) causes the corners to be squared off with straight slices.

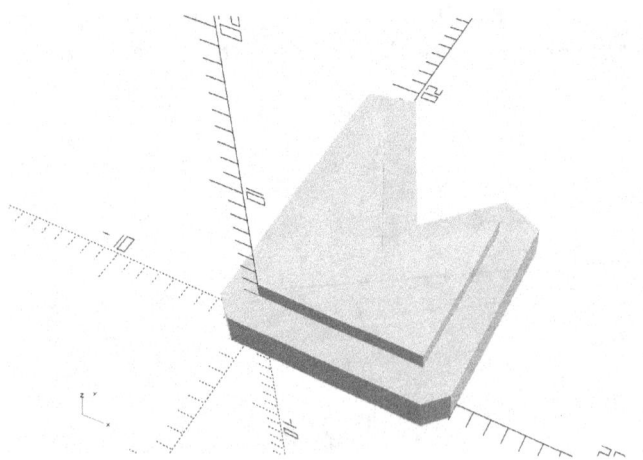

When the first parameter is set to a negative value, the new offset outline shifts inwards from the original shape. I've changed the height parameters to make the resulting offsets higher than the original polygon, keeping both in view.

Here's the result of applying offset_round(-1,4). The only round part is at the indentation corner near the center of the offset outline.

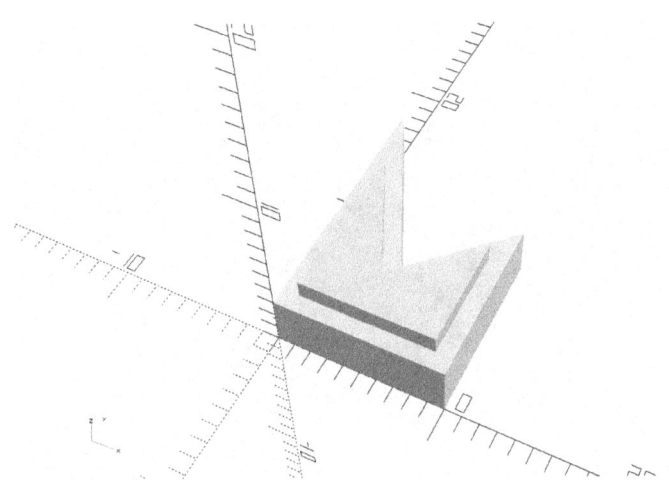

offset_straight(-1,4) again creates all straight corner lines in the taller inset shape that's recessed inside the original polygon.

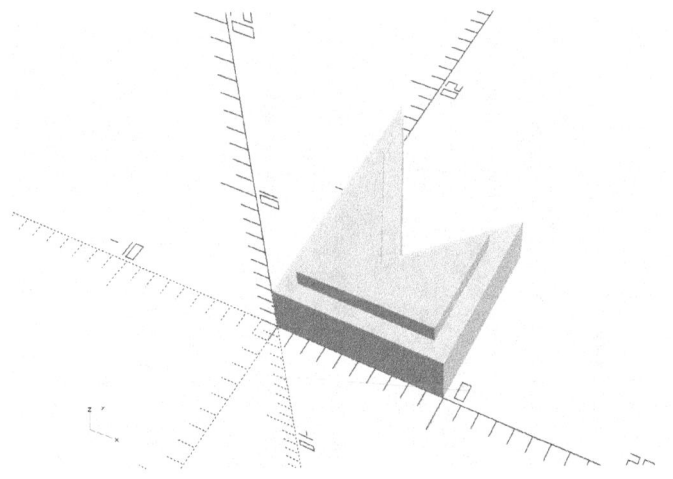

And finally, the offset_chamfer(-1,4) causes the internal corner to be squared off.

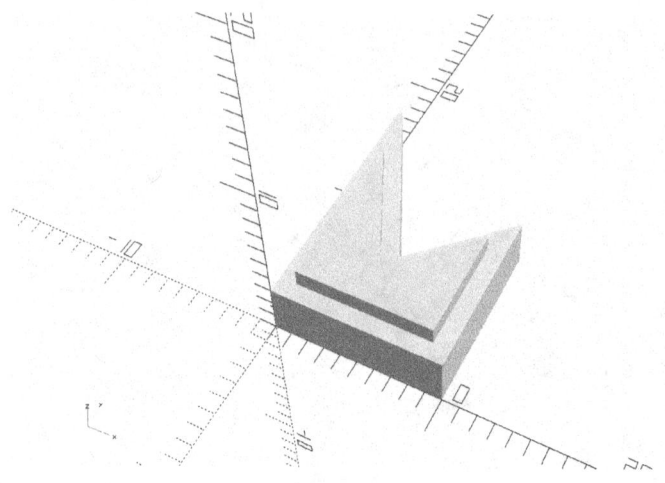

24 - Difference, Union, & Intersection

The difference(), union(), and intersection() commands all work the same way, although the results are different. We can create multiple objects, then add them or subtract them together in space in these various ways. For example, earlier we created a hollowed out cone by using the difference() command to subtract an inner cone from an outer one.

When two or more objects are created in space in an overlapping way, the default if we do nothing is for them to be combined using union(). If you export such a construction to an STL file, the visible surface of the whole combined set of objects is calculated, and none of the internal "surfaces" are used. The union() command kind of melts all the shapes into one glued together whole.

The difference() command starts with the first object listed, and vaporizes all the rest of them. So in a list of two or more objects, the difference() command leaves only the parts of the first object where none of the remaining objects existed.

The intersection() command also erases parts of objects from existence, but only those parts where they don't overlap with the first part in the list.

Let's create a sphere and a box that coexist in an overlapping way in space. We'll then proceed to apply the various commands just mentioned to see what happens.

```
import openscad as o

# Create a cube
o.box(10,10,10)
box = o.result()

# Create an intersecting sphere
o.sphere(10)
o.translate(10,5,10)
sphere = o.result()

# Send the results to OpenSCAD
o.output([box,sphere])
```

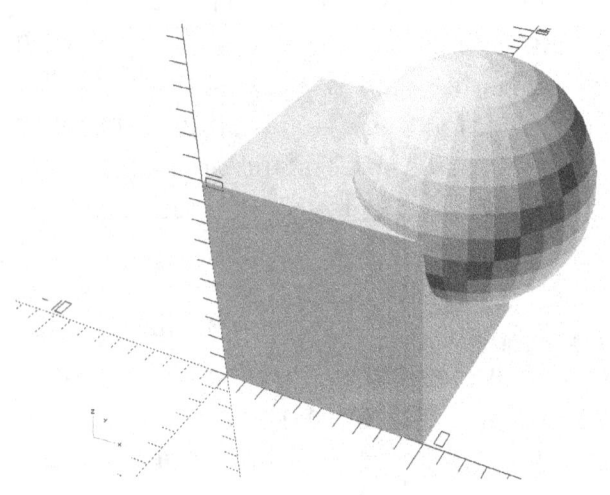

As mentioned before, the default action is a union between these two objects, and an export to STL file of the results as they stand would create exactly what you see here.

To demonstrate the equivalence of the union() command, the following code creates exactly the same output.

```
import openscad as o

# Create a cube
o.box(10,10,10)
box = o.result()

# Create an intersecting sphere
o.sphere(10)
o.translate(10,5,10)
sphere = o.result()

o.union([box,sphere])

# Send the results to OpenSCAD
combination = o.result()
o.output(combination)
```

Each of these commands can be passed a list of more than two objects if desired. The difference() command subtracts all the objects after the first one in the list, from the first one in the list. With our sphere and box, we can set this up two ways, with unique results in each case. First, let's list the box first, and this causes the sphere to be erased from space, carving a big bite out of the box.

```python
import openscad as o

# Create a cube
o.box(10,10,10)
box = o.result()

# Create an intersecting sphere
o.sphere(10)
o.translate(10,5,10)
sphere = o.result()

o.difference([box,sphere])

# Send the results to OpenSCAD
combination = o.result()
o.output(combination)
```

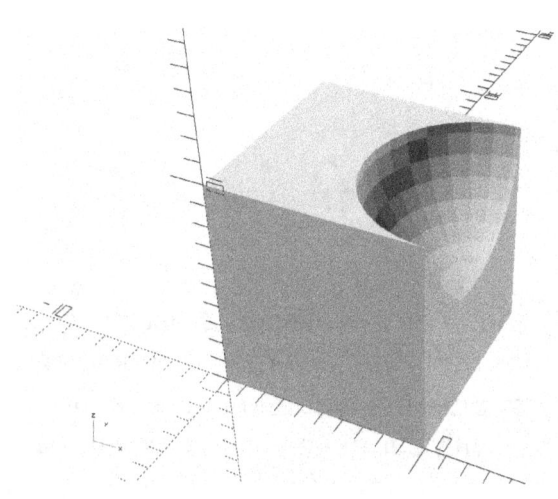

If we reverse the order of the objects in the difference list, so the sphere comes first, we end up with a hunk of the box removed from the sphere.

```python
import openscad as o

# Create a cube
o.box(10,10,10)
box = o.result()

# Create an intersecting sphere
o.sphere(10)
o.translate(10,5,10)
sphere = o.result()

o.difference([sphere,box])

# Send the results to OpenSCAD
combination = o.result()
o.output(combination)
```

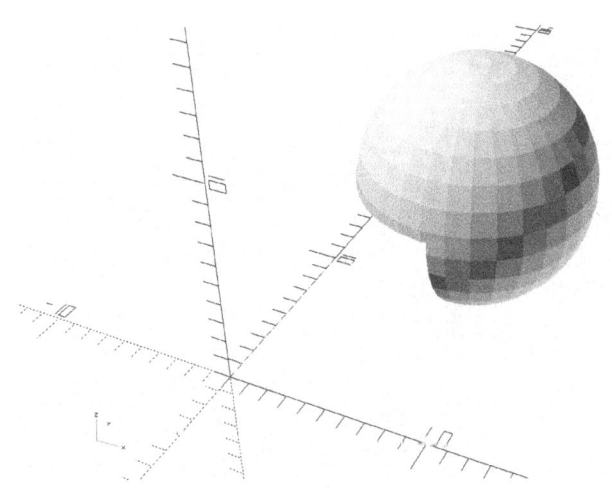

The intersection() command erases everything except those parts where shapes overlap. In this case, our box and sphere intersect in a small, kind of lemon-wedge shaped volume they share in space.

```python
import openscad as o

# Create a cube
o.box(10,10,10)
box = o.result()

# Create an intersecting sphere
o.sphere(10)
o.translate(10,5,10)
sphere = o.result()

o.intersection([box,sphere])

# Send the results to OpenSCAD
combination = o.result()
o.output(combination)
```

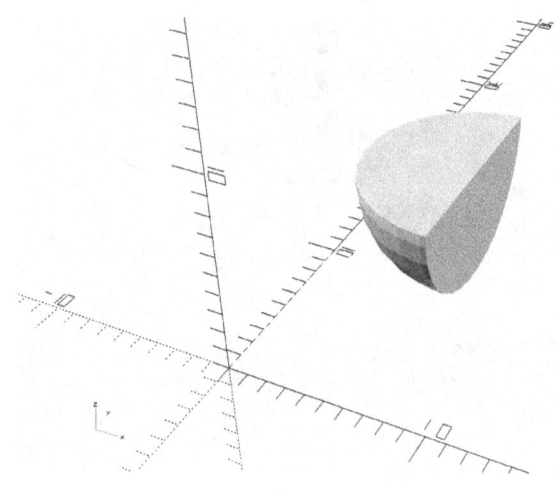

Keep in mind that these examples created lists containing only two objects, the box and the sphere. Longer lists of objects work fine too, and the basic rule to remember is that the commands operate using all objects from the second one on against the first object in the list.

25 - Assemblies

As your projects grow in complexity and size, a good way to modularize things is to place objects in their own files, and then create a central project that pulls those files in to create a complete assembly. This chapter demonstrates this technique with a straightforward example where a bolt, a nut, and a washer are created in their own files, and they are then pulled together into a complete assembly.

First, let's create the individual parts and save each construction in its own file. Here's the code to be saved in the first file, named nut.py.

```
from openscad import *

# Main shape of nut
regular_polygon(6,7,3)
nut_parts = [result()]

# Define hole through nut
cyl(6,5)
translate(0,0,-1)
nut_parts.append(result())

# Subtract out the hole
difference(nut_parts)
translate(0,0,22)
rotate(0,0,30)
color("Silver")

nut = result()
```

If you try to run this Python code, nothing really happens. The nut object is created okay, but it doesn't get written into an OpenSCAD file with an extension of ".scad". However, it is ready to be included into a bigger project, along with the other parts.

Next, create the file bolt.py containing the following code.

```
import openscad as o
o.fragments = 90

# Head of the bolt
o.regular_polygon(6,7,3)
o.color("Silver")

# Rod part of the bolt
o.cyl(6,19)
o.translate(0,0,1)
o.color("DarkGray")

bolt = o.result()
```

Now, create the file that creates the washer, appropriately named washer.py.

```
from openscad import *

tube(16,7,1)
translate(0,0,9)
color("LightSlateGray")

washer = result()
```

Finally, create our main project file, and name it assembly.py.

```
#assembly.py

import openscad
from bolt import *
from washer import *
from nut import *

assembly_list = [bolt,washer,nut]
o.output(assembly_list)
```

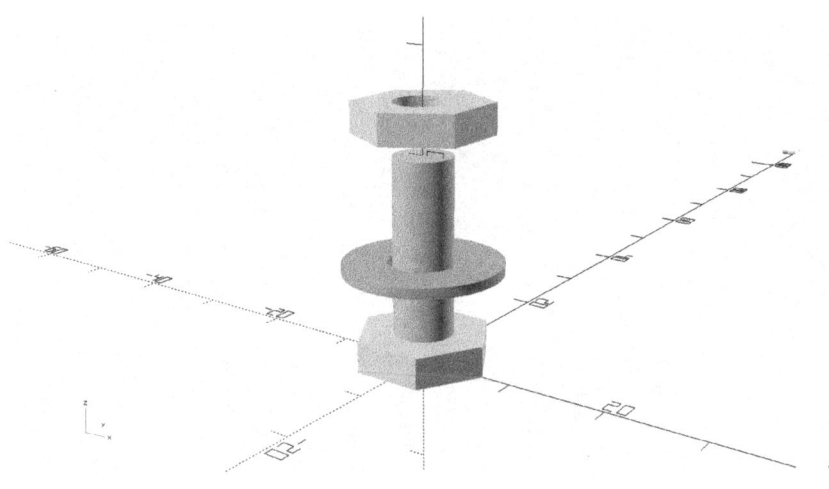

This main project file is easy to read. Each file is imported using Python's "from file import *" statement, which allows the variables named in the files, such as bolt, washer, and nut, to be referenced directly in the list named assembly_list.

You can also use other standard Python techniques to place functions and other code in these files. The simple example presented here should provide a big head start on how to modularize your projects as they grow bigger.

26 - Gears

As your projects get more technically complicated, the true power and awesomeness of Python starts to shine through. This chapter presents a significant module that creates an N-toothed involute gear. I created this code after doing a lot of research on the Internet, learning that I really didn't know what I didn't know about gears at first. If you're curious, or if you plan to create and use a lot of gears in your projects, I strongly suggest you do some research yourself to get up to speed with the math and logic behind creating smoothly meshing involute-teeth gears. There's a lot of cool science and math to learn here!

The file named gear.py contains the main gear() function, along with several supporting functions it in turn calls. The file is all functions, and they all get loaded before anything gets called. At the end of the listing is the standard Python way to start the action by calling the main() function, assuming this file is run stand-alone. I set it up this way so the gear() function can be called by another program that loads this file. We'll do that next, by creating a cool program that creates two gears and then animates them.

First though, here's the complete code listing for the gear.py file.

Python for 3D Printing

```python
# gear
# http://bit.ly/2VxeUCA

import os
import math
import openscad as o


def main():
    gear(16,2,14.5,0.75,1.0)
    o.output(o.result())




def gear(N,P,PA,GT,HD):
    # N    Number of teeth      12
    # P    Diameteral pitch      2
    # PA   Pressure angle       14.5
    # GT   Gear thickness        0.75
    # HD   Center hole diameter  1.0

    o.fragments = 100
    aFrags = 12

    PR = N / (2 * P)                         # Pitch circle radius
    d = 1.157 / P                            # Dedendum
    OR = (N + 2) / (2 * P)                   # Outside circle radius
    RR = PR - d                              # Root circle radius
    BR = PR * math.cos(math.radians(PA))     # Base circle radius
    TL = (OR ** 2 - BR ** 2) ** 0.5          # Tangent length
    AL = TL                                  # Arc length
    AA = AL / BR                             # arc = theta * radius
    FR = 0.3 / P                             # Root circle fillet radius
    teeth = []
```

```python
# Create notch polygons between each pair of teeth
for tooth in range(N):

    # Create 2D outline of notch shape as list of points
    notch_outline = []

    # Tooth contact involute curve points
    for i in range(aFrags):
        j = i + 1
        dist = TL * j / aFrags
        ang = AA * j / aFrags
        p1 = [0,0]
        p2 = [BR * math.cos(ang), BR * math.sin(ang)]
        p3 = d2_turn_right(p1,p2,dist)
        p4 = d2_rotate(p3,90 / N)
        p5 = [p4[0],p4[1]]
        notch_outline.append(p5)

    # Outer outline dot
    p1 = [2 * OR - BR,0]
    notch_outline.append(p1)

    # Tooth contact opposite involute curve
    for i in range(aFrags):
        j = aFrags - i
        dist = TL * j / aFrags
        ang = AA * j / aFrags
        p1 = [0,0]
        p2 = [BR * math.cos(ang), BR * math.sin(ang)]
        p3 = d2_turn_right(p1,p2,dist)
        p4 = d2_rotate(p3,90 / N)
        p5 = [p4[0],-p4[1]]
        notch_outline.append(p5)

    # Root circle plus dot below center
    p1 = [RR + FR,0]
    p2 = d2_rotate(p1, -90 / N)
    notch_outline.append(p2)
```

```python
    # Root circle line dot below center
    p1 = [RR,FR]
    p2 = d2_rotate(p1, -90 / N)
    notch_outline.append(p2)

    # Root circle line dot above center
    p1 = [RR,-FR]
    p2 = d2_rotate(p1, 90 / N)
    notch_outline.append(p2)

    # Root circle plus dot above center
    p1 = [RR + FR,0]
    p2 = d2_rotate(p1, 90 / N)
    notch_outline.append(p2)

    # Create 3D cutout notch between teeth
    o.polygon(notch_outline,GT + 2)
    o.translate(0,0,-1)
    o.rotate(0,0,360 * tooth / N)
    notch = o.result()

    # Add to list of notches
    teeth.append(notch)

# Create the main round gear disk
o.cyl(OR+OR,GT)
gear_disk = o.result()

# Prepend the disk to subtract notches
teeth.insert(0,gear_disk)



# Append the center hole to subtract it too
o.cyl(HD,GT + 2 )
o.translate(0,0,-1)
center_hole = o.result()
teeth.append(center_hole)

# Subtract notches and hole from gear disk
o.difference(teeth)

# Optional: Add two pitch radius values to get
# center to center distance for two meshed gears
print("Pitch radius: ",PR)
```

```
def d2_turn_left(p1, p2, dist):
    v = d2_unit([p2[0] - p1[0], p2[1] - p1[1]])
    vleft = [-v[1], v[0]]
    p3 = []
    p3.append(p2[0] + dist * vleft[0])
    p3.append(p2[1] + dist * vleft[1])
    return p3


def d2_turn_right(p1, p2, dist):
    return d2_turn_left(p1, p2, -dist)


def d2_rotate(p, deg):
    rad = math.radians(deg)
    x = p[0] * math.cos(rad) - p[1] * math.sin(rad)
    y = p[1] * math.cos(rad) + p[0] * math.sin(rad)
    return [x, y]


def d2_mag(v):
    return (v[0] ** 2 + v[1] ** 2) ** 0.5


def d2_unit(v):
    mag = d2_mag(v)
    return [v[0] / mag, v[1] / mag]


if __name__ == '__main__':
    main()
```

When you run this Python program, this is what shows up in the OpenSCAD window.

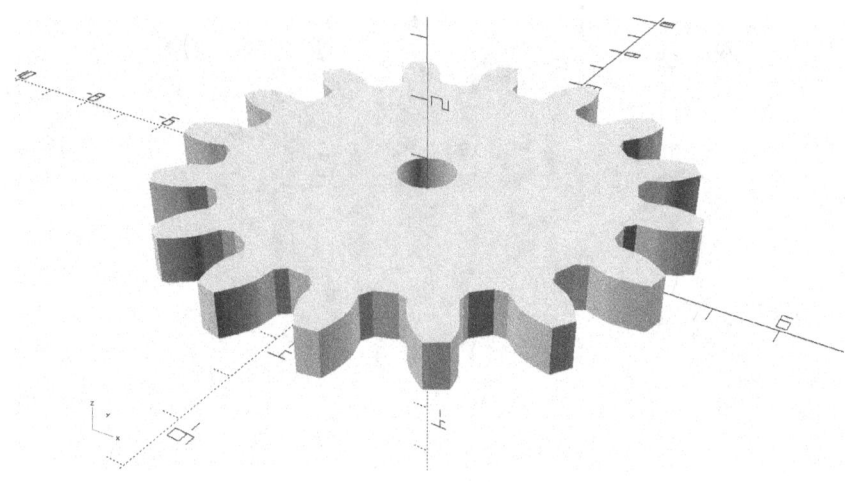

At the top of the file, in the main() function, you can change the various parameters in the call to gear(), to change the number of teeth, the diametral pitch, and so on.

Also note when the program runs that a Python print() function outputs the "Pitch radius" of the created gear. This is the distance from the center of the gear to its pitch circle. This number is handy for spacing gears such that they mesh correctly.

Animated Gears

This next example creates two gears by importing the gear.py file and calling gear() twice, once to create a gear with 9 teeth, and again to create a second gear with 16 teeth. The two gears are translated in different directions away from the origin based on their two pitch radii, and they are rotated to align their teeth correctly for proper meshing. The $t string variable was used in each gear's rotation calculation, allowing animation once everything is said and done. Here's the code, to be saved in gears_animation.py.

```python
# gears_animation
# http://bit.ly/2VxeUCA
# Animate FPS: 30 Steps: 30

import os
import math
import openscad as o
import gear

# Create first gear
teeth1 = 9
gear.gear(teeth1,2,14.5,0.75,1.0)

# Rotate to mesh teeth correctly
o.rotate(0,0,f"-360 * $t/{teeth1}")

# Move to the right by pitch radius
o.translate(2.25,0,0)
gear1 = o.result()

# Create second gear
teeth2 = 16
gear.gear(teeth2,2,14.5,0.75,1.0)

# Rotate to mesh teeth correctly
o.rotate(0,0,f"360 * $t/{teeth2}")

# move to the left by pitch radius
o.translate(-4.0,0,0)
gear2 = o.result()

# Send result to OpenSCAD
o.output([gear1,gear2])
```

To view the animation, from OpenSCAD's View menu select Animate. Try setting both FPS and Steps to 30, and adjust these as desired.

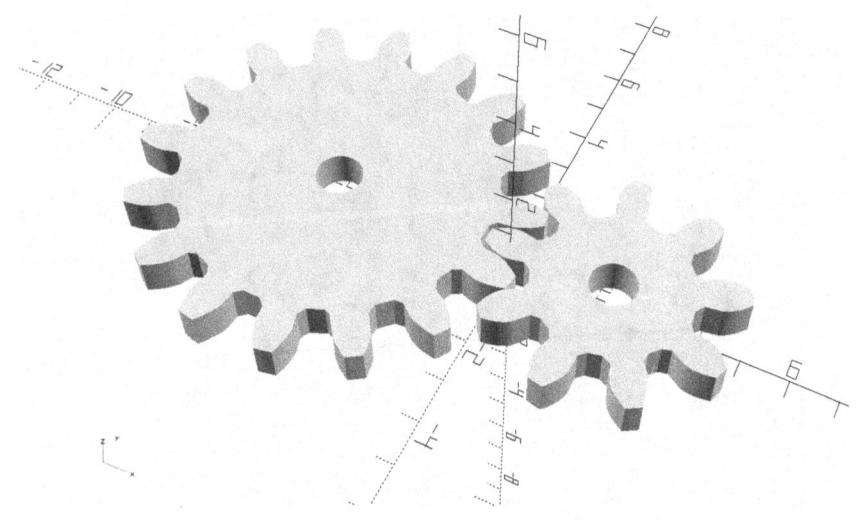

27 - Mason Bees

Mason bees? In a book called Python for OpenSCAD?

Yep, this chapter presents a fun and easy project to create homes for Mason bees. If you haven't heard about Mason bees, they are awesome. They pollinate vegetables and other plants about 100 times more efficiently than honeybees, they don't sting, and they are fun and easy to support near your home and garden. They are fascinating, and very eco-friendly.

You can find instructions online for creating Mason bee homes using wood or similar materials, or you can buy them already made. Basically, long holes drilled in wood, or tubes such as hollow bamboo are what these bees like for setting up camp. Cocoons for drones and queens are stored in these tubes, ready for the start of each season. All you need to do is supply the holes, and the queen bees will know what to do with them.

I've set up the following Python code in a highly parameterized way, so you can change the values for certain variables at the start of the program to control the end results. The holes are created in a block just under 6 inches long, so it should fit in the printing volume of many 3D printers. The number of rows and columns can be adjusted, to make anything from the simplest 1-hole tube up to a much larger set, such as 16 or 25 holes. I suggest starting at about 8 holes. Here's the code, with parameters set for a typical Mason bee home with 16 holes in a 4 x 4 array.

```python
# mason_bees.py
from openscad import *

# All units are mm
holes_per_row = 4
hole_diameter = 9.4
separation = 14
box_height = 150
back_wall = 7

# Calculate some dimensions
boxsize = separation * (holes_per_row + 1) - hole_diameter
offset = boxsize / 2 - separation * (holes_per_row - 1) / 2

# Create the tall rectangular solid
box(boxsize, boxsize, box_height)
box_parts = [result()]

# Add array of holes
for x in range(holes_per_row):
    for y in range(holes_per_row):
        cyl(hole_diameter, box_height)
        hole_x = x * separation + offset
        hole_y = y * separation + offset
        translate(hole_x, hole_y, back_wall)
        box_parts.append(result())

# Subtract the holes from the box
difference(box_parts)
bee_box = result()
output(bee_box)
```

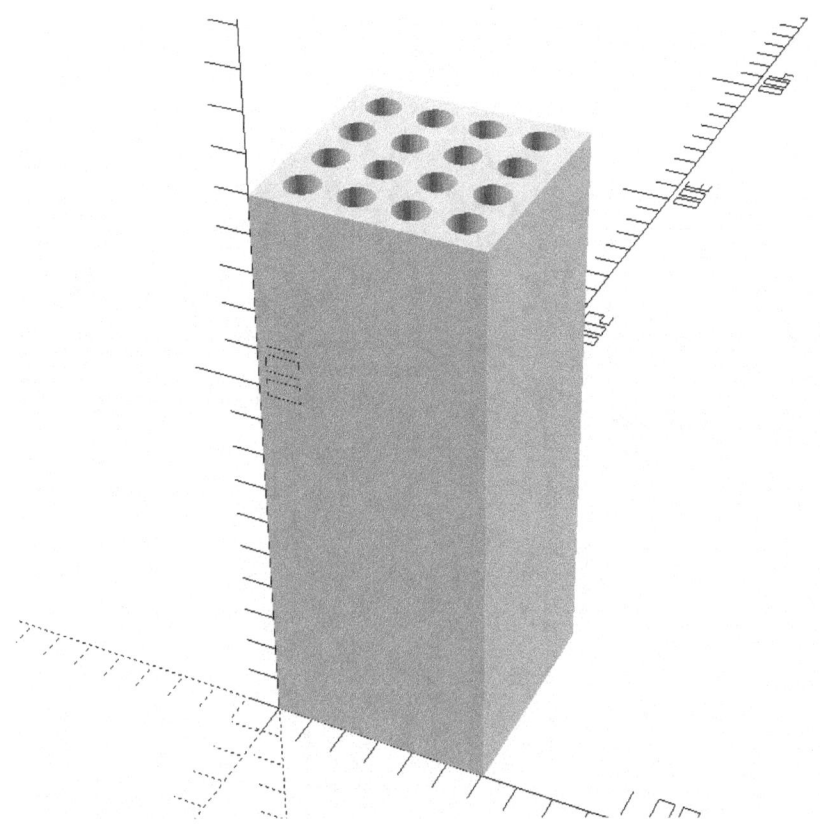

I would strongly suggest lining these holes with parchment paper, as described several places online. This should provide the right "look and feel" for the bees to make their home, and it can provide a way to clean out the holes each season.

28 - Surface

The surface() command loads either an image file, or a specially constructed data file, and creates a 3D height map from the pixels after they are turned to gray scale. The pixel heights are scaled from 0 to 100, based on their gray scale value.

The following example loads a photo of my wife and creates a surface map. I think she's much more beautiful in real 3D of course. The scale() command brings the range of the height of the pixels from 0 to 100 down into the range 0 to 3, while leaving the width and height of the result the same as the number of pixels in the image file.

```
import openscad as o

o.surface("L256.png")
o.scale(1,1,.03)

o.output(o.result())
```

It's also easy to construct a text file containing space-separated integers, where each line of text represents a row of pixels. Here's some example code that creates a file named example.dat, fills it with integers calculated using a SIN(x)/SIN(x) function, and then calls surface() to display the data as a surface map.

```
import openscad as o
import math

f = open("example.dat","w")
size = 100
for i in range(size):
    for j in range(size):
        x = (i-size/2) / 3
        y = (j-size/2) / 3
        d = (x*x + y*y)**.5
        if d:
            h = math.sin(d)/d
        else:
            h = 1
        f.write(f"{int(h * 100)} ")
    f.write("\n")
f.close

o.surface("example.dat")
o.scale(1,1,.2)

o.output(o.result())
```

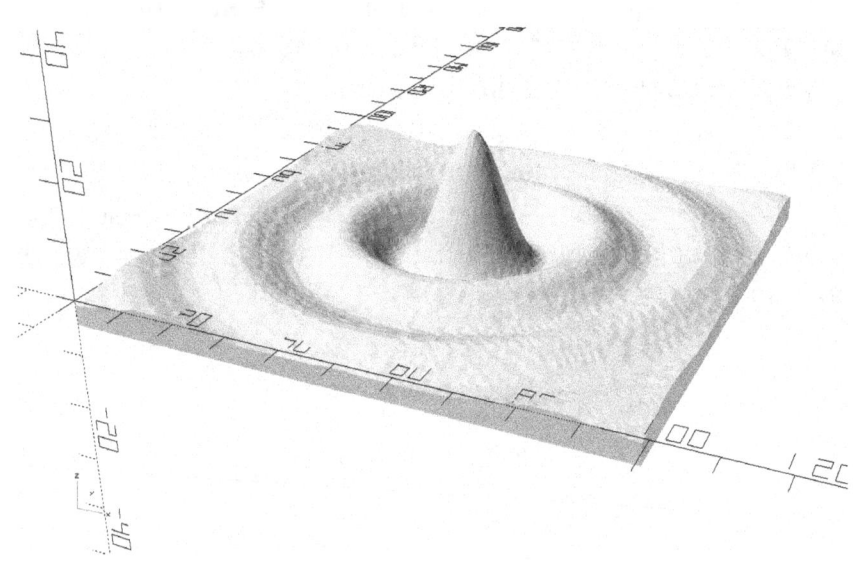

If you zoom in close on the photo display, you'll see the surface is comprised of polygons, with edges connecting in pointed peaks and troughs.

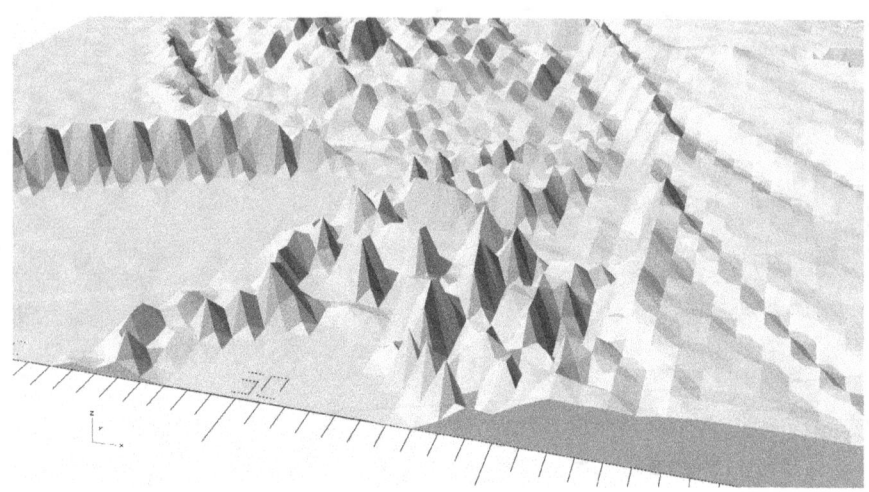

As an experiment I decided to try my hand at using Python's image processing library, named pillow, to create a slightly different kind of surface map. My algorithm converts the image to grayscale, then creates many small boxes where the height of each box is proportional to the pixel intensity at that X, Y location. The results are similar, although you can really see the difference when you zoom in, as shown in the following example.

```python
import openscad as o
from PIL import Image
from PIL import ImageFilter
from PIL import ImageOps

siz = 128
grayscale = 4
base = 4

# Open pretty much any image file
fil = 'L256.png'
img = Image.open(fil)

# Resize to the desired size
img = img.resize((siz,siz),Image.ANTIALIAS)

# Process for edge detection or other filters
img = img.filter(ImageFilter.EDGE_ENHANCE)

# Convert to 256 level grayscale
img = img.convert('L')

# Convert to grayscale level less than 256
pix = img.load()
for x in range(siz):
    for y in range(siz):
        pix[x,y] = int(pix[x,y] / (256/grayscale))

# Create a working "array" for the pixels
res = [[0 for i in range(siz)] for j in range(siz)]

# Process the pixels
pix = img.load()
for x in range(siz):
    for y in range(siz):
        res[x][y] = pix[x,y] + base

# Create combined output boxes for efficiency
for x in range(siz):
    y1 = 0
    h = res[x][0]
    for y in range(siz):
        if (res[x][y] != h) or (y == siz - 1):
            o.box(1,y-y1,h)
            o.translate(x,-y,0)
            h = res[x][y]
            y1=y

# All done
o.output(o.result())
```

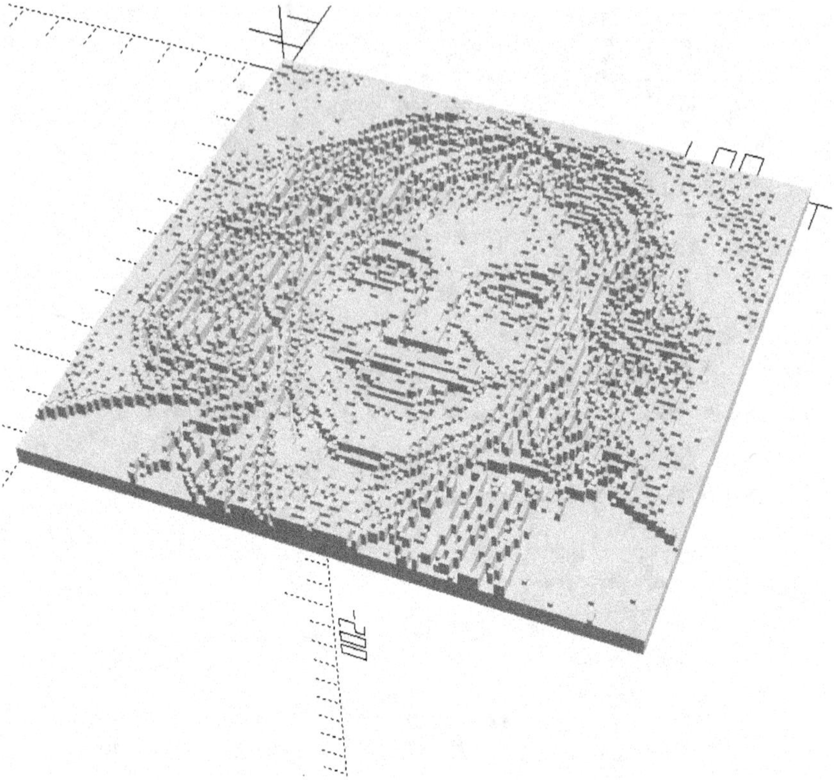

Python for 3D Printing

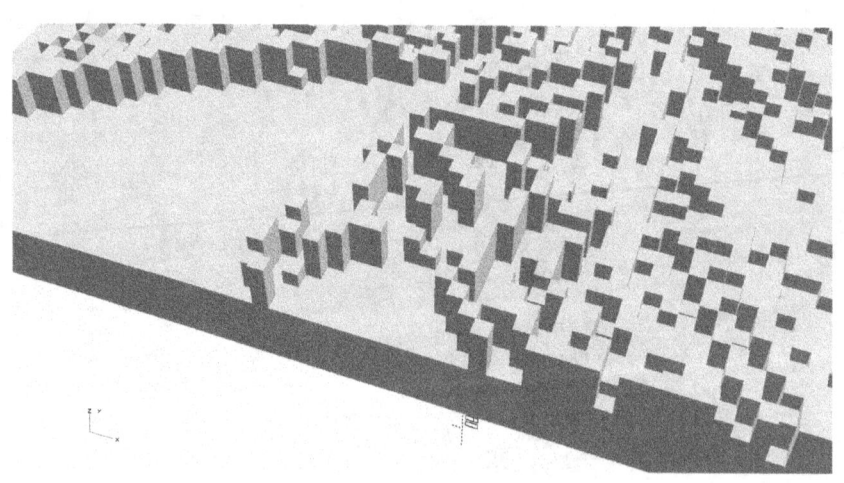

29 - Platonic solids

One feature missing from OpenSCAD is the ability to create most of the regular platonic solids. The cube is easy, but the other four are not so easy.

This chapter presents Python code to create all five platonic solids. In addition to the cube, they include tetrahedron, octahedron, dodecahedron, and icosahedron. The one parameter in each case is the edge length. In all the examples to follow, the edge lengths are set to 10 units.

To use these in your own projects, you'll perhaps need to adjust the edge lengths and their orientation in space. This code will at least give you a big head start.

Tetrahedron

```
from openscad import *

platonic_tetrahedron(10)

# Send result to OpenSCAD
output(result())
```

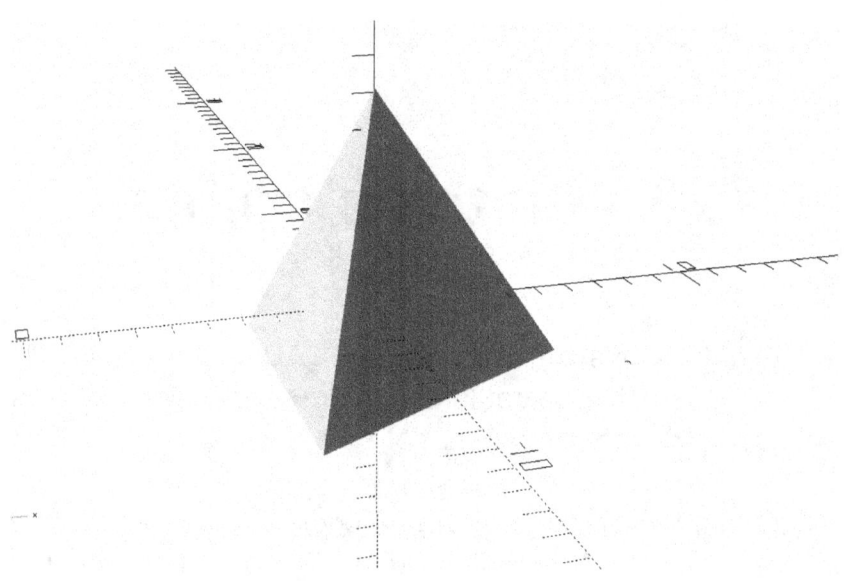

Cube

```
from openscad import *

platonic_cube(10)

# Send result to OpenSCAD
output(result())
```

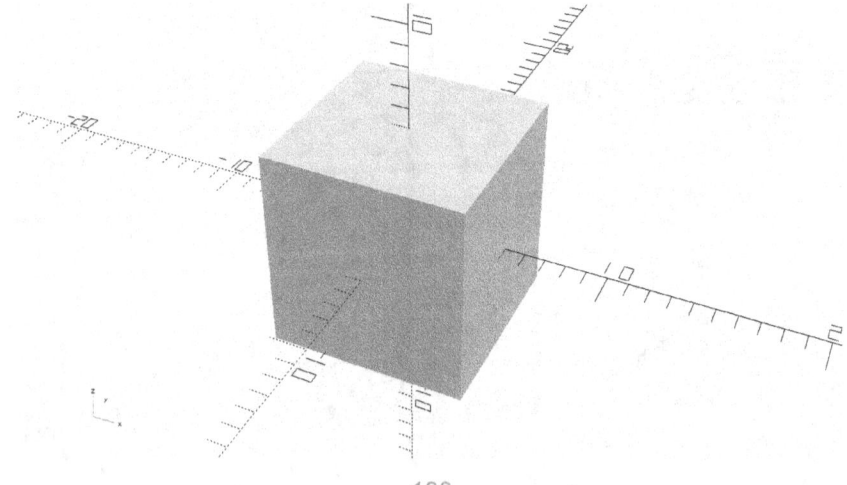

Octahedron

```
from openscad import *

platonic_octahedron(10)

# Send result to OpenSCAD
output(result())
```

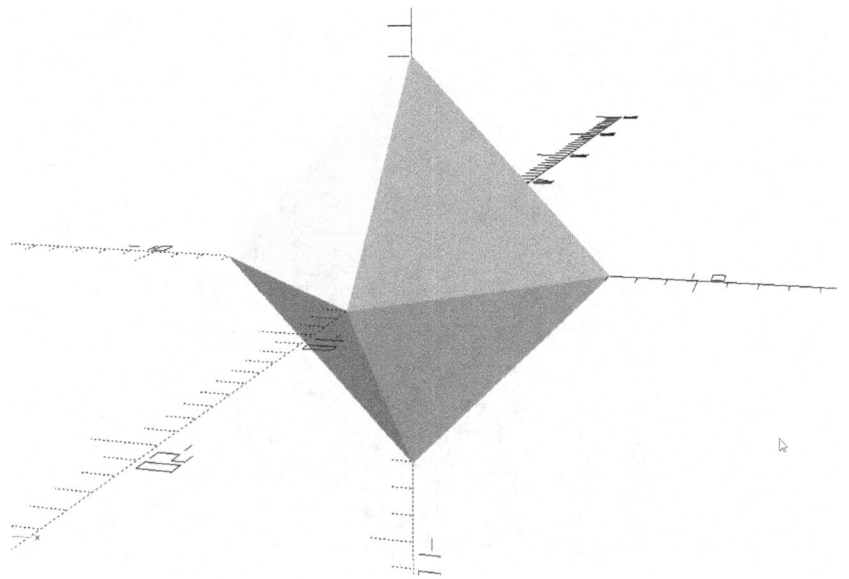

Dodecahedron

```
from openscad import *

platonic_dodecahedron(10)

# Send result to OpenSCAD
output(result())
```

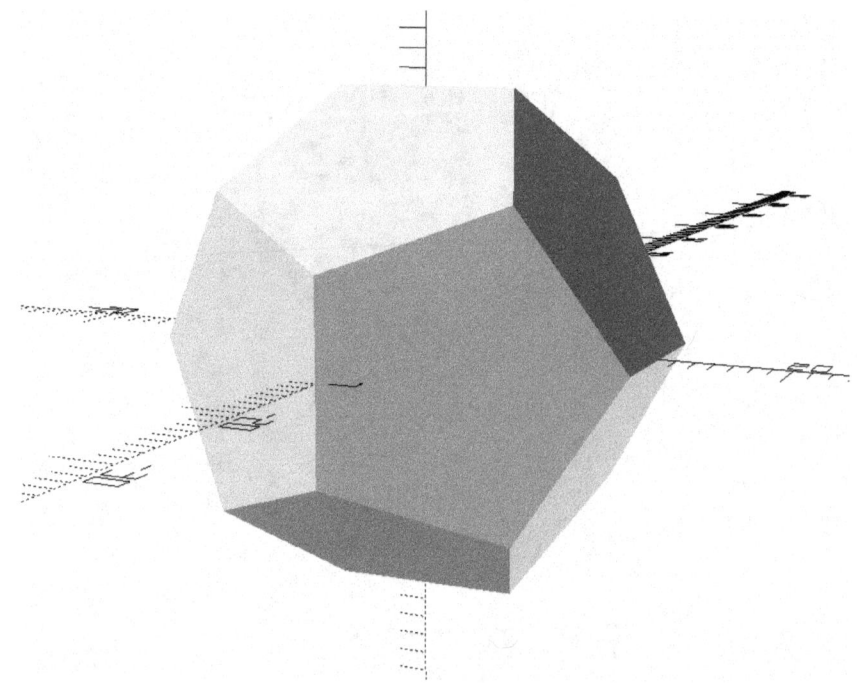

Icosahedron

```
from openscad import *

platonic_icosahedron(10)

# Send result to OpenSCAD
output(result())
```

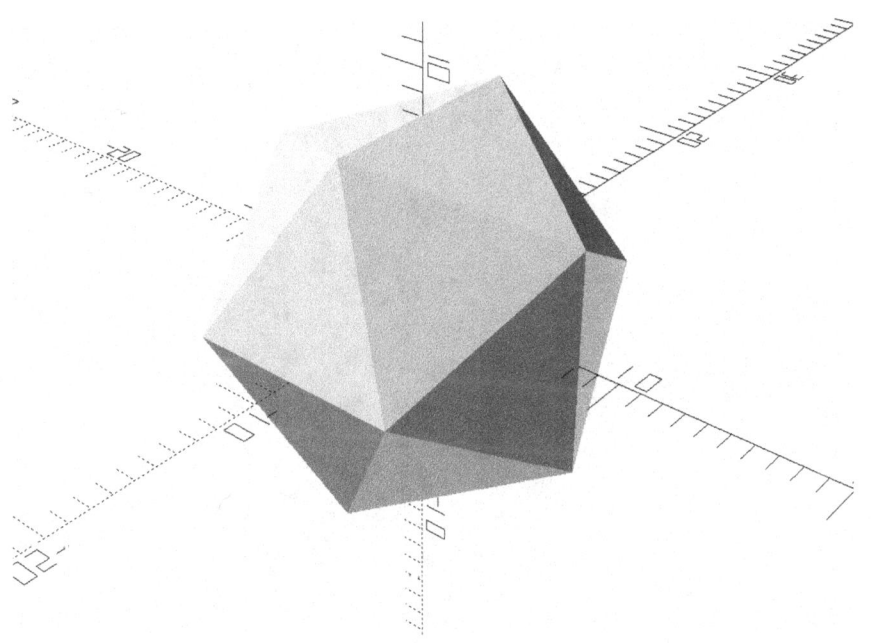

Appendix A

openscad.py source code

Here's the source code for the file openscad.py. Place a copy of this file in any project folder where you want to use the Python for OpenSCAD technique presented throughout this book.

I'll be glad to email you a copy of this file if you send me a short request...john@openscadbook.com

```python
# openscad.py

import os
import math
import inspect

# Global commands accumulator
_cmds = ""
_use = ""

# Function modifying parameters
fragments = 31

# Constants
tiny = 1e-99
```

```python
# Functions for driving Openscad


def startup(s):
    global _use
    _use += f"""{s}\n"""


def literally(s):
    global _cmds
    _cmds += f"""{s}\n"""


def cyl(diameter, height):
    global _cmds, fragments
    radius = diameter / 2
    _cmds = (
        f"cylinder(h={height},"
        f"r1={radius},r2={radius},"
        f"center=false,$fn={fragments});\n\n"
    ) + _cmds


def cone(diameter, height):
    global _cmds, fragments
    radius = diameter / 2
    _cmds = (
        f"cylinder(h={height},"
        f"r1={radius},r2={0},"
        f"center=false,"
        f"$fn={fragments});\n\n"
    ) + _cmds
```

```python
def cone_truncated(diameter1, diameter2, height):
    global _cmds, fragments
    radius = diameter1 / 2
    radius2 = diameter2 / 2
    _cmds = (
        f"cylinder(h={height},r1={radius},"
        f"r2={radius2},center=false,"
        f"$fn={fragments});\n\n"
    ) + _cmds


def sphere(diameter):
    global _cmds, fragments
    radius = diameter / 2
    _cmds = (f"sphere({radius},"
        f"$fn={fragments});\n\n" ) + _cmds


def box(x, y, z):
    global _cmds
    _cmds = (f"cube({[x,y,z]},"
        f"center=false);\n\n") + _cmds


def polygon(points_list, height):
    global _cmds, fragments
    _cmds = (
        f"linear_extrude({height})\n"
        f"polygon({points_list},"
        f"convexity=20,$fn={fragments});\n\n"
        ) + _cmds
```

```python
def triangle(point1, point2, point3, height):
    polygon([point1, point2, point3], height)


def regular_polygon(sides, radius, height):
    global _cmds
    _cmds = "}\n\n" + _cmds
    for wedge in range(sides):
        p1 = _cart(radius,wedge*360/sides)
        p2 = _cart(radius,(wedge+1)*360/sides)
        triangle([0, 0], p1, p2, height)
    _cmds = "union(){\n" + _cmds


def tube(outside_diam, inside_diam, height):
    global _cmds, fragments
    r1 = outside_diam / 2
    r2 = inside_diam / 2
    _cmds = (
        "difference(){\n"
        f"cylinder(h={height},r1={r1},r2={r1},"
        f"center=false,$fn={fragments});\n"
        f"cylinder(h={height*3},r1={r2},r2={r2},"
        f"center=true,$fn={fragments});\n"
        "}\n" ) + _cmds


def polyhedron(points_list, faces_list):
    global _cmds
    _cmds = (f"polyhedron({points_list},"
        f"{faces_list},convexity=20);\n\n"
        ) + _cmds
```

```python
def text(
    text="",
    size=10,
    font="Liberation Sans",
    halign="left",
    valign="baseline",
    spacing=1,
    direction="ltr",
    language="en",
    script="latin",
    height=1,
):
    global _cmds, fragments
    _cmds = (
        f'linear_extrude({height})\n'
        f'text(text="{text}",size={size},'
        f'font="{font}",halign="{halign}",'
        f'valign="{valign}",spacing={spacing},'
        f'direction="{direction}",'
        f'language="{language}",'
        f'script="{script}",'
        f'$fn={fragments});\n\n'
    ) + _cmds


def translate(x, y, z):
    global _cmds
    _cmds = f"translate([{x},{y},{z}])\n" + _cmds


def rotate(x, y, z):
    global _cmds
    _cmds = f"rotate([{x},{y},{z}])\n" + _cmds


def scale(x, y, z):
    global _cmds
    _cmds = f"scale([{x},{y},{z}])\n" + _cmds
```

```python
def resize(x, y, z):
    global _cmds
    _cmds = f"resize([{x},{y},{z}])\n" + _cmds


def mirror(x, y, z):
    global _cmds
    _cmds = f"mirror([{x},{y},{z}])\n" + _cmds


def color(color_name, alpha=1.0):
    global _cmds
    _cmds = (f'color("{color_name}",'
        f'{alpha})\n') + _cmds


def rgb(r, g, b, alpha=1.0):
    global _cmds
    _cmds = (f"color([{r/255},{g/255},"
        f"{b/255},{alpha}])\n") + _cmds


def offset_round(distance, height):
    global _cmds, fragments
    _cmds = (
        f"linear_extrude({height})\n"
        f"offset(r={distance},chamfer=false,"
        f"$fn={fragments})\n"
        f"projection()\n" ) + _cmds
```

```python
def offset_straight(distance, height):
    global _cmds
    _cmds = (
        f"linear_extrude({height})\n"
        f"offset(delta={distance},"
        f"chamfer=false)\n"
        f"projection()\n"
    ) + _cmds


def offset_chamfer(distance, height):
    global _cmds
    _cmds = (
        f"linear_extrude({height})\n"
        f"offset(delta={distance},"
        f"chamfer=true)\n"
        f"projection()\n"
    ) + _cmds


def projection(height):
    global _cmds
    _cmds = (f"linear_extrude({height})"
        f"\nprojection()\n") + _cmds


def slice(height):
    global _cmds
    _cmds = (f"linear_extrude({height})\n"
        f"projection(cut=true)\n") + _cmds
```

```python
def spiral(turns, height, scale=1):
    global _cmds, fragments
    _cmds = (
        f"linear_extrude(height={height},"
        f"twist=-{360*turns},"
        f"scale={scale},"
        f"slices={fragments},"
        f"center=false,"
        f"convexity=20,"
        f"$fn={fragments})\n"
        f"projection()\n"
    ) + _cmds


def rotate_extrude(angle=360):
    global _cmds, fragments
    _cmds = (
        f"rotate_extrude(angle={angle},"
        f"convexity=20,$fn={fragments})\n"
        f"projection()\n"
    ) + _cmds


def union(obj_list):
    global _cmds
    cmd = "union(){\n"
    for obj in obj_list:
        cmd += obj
    cmd += "}\n"
    _cmds = cmd + _cmds
```

```python
def difference(obj_list):
    global _cmds
    cmd = "difference(){\n"
    for obj in obj_list:
        cmd += obj
    cmd += "}\n"
    _cmds = cmd + _cmds


def intersection(obj_list):
    global _cmds
    cmd = "intersection(){\n"
    for obj in obj_list:
        cmd += obj
    cmd += "}\n"
    _cmds = cmd + _cmds


def hull(obj_list):
    global _cmds
    cmd = "hull(){\n"
    for obj in obj_list:
        cmd += obj
    cmd += "}\n"
    _cmds = cmd + _cmds


def minkowski(obj_list):
    global _cmds
    cmd = "minkowski(){\n"
    for obj in obj_list:
        cmd += obj
    cmd += "}\n"
    _cmds = cmd + _cmds
```

```python
def surface(filename,invert="false"):
    global _cmds
    _cmds = (f'surface(file=\"{filename}\",'
        f'convexity=5,'
        f'invert={invert});\n') + _cmds


def result():
    global _cmds
    res = _cmds
    _cmds = ""
    return res


def output(cmds_list=""):
    global _use
    calling_file = inspect.stack()[1].filename
    srcfile = calling_file.split("\\")[-1]
    dstfile = srcfile.replace(".py", ".scad")
    f = open(dstfile, "w")
    if _use:
        f.write(_use)
        _use = ""
    for cmd in cmds_list:
        f.write(cmd)
    f.close()


def platonic_cube(n):
    box(n, n, n)
    translate(-n / 2, -n / 2, -n / 2)
```

```python
def platonic_tetrahedron(n):
    n0 = n * (8 / 3) ** -0.5
    n1 = ((8 / 9) ** 0.5) * n0
    n2 = ((2 / 9) ** 0.5) * n0
    n3 = ((2 / 3) ** 0.5) * n0
    n4 = -n0 / 3
    v0 = [n1, 0, n4]
    v1 = [-n2, n3, n4]
    v2 = [-n2, -n3, n4]
    v3 = [0, 0, n0]
    points = [v0, v1, v2, v3]
    faces = [[0, 1, 2], [0, 3, 1],
             [1, 3, 2], [2, 3, 0]]
    polyhedron(points, faces)


def platonic_octahedron(n):
    n0 = n * 2 ** -0.5
    v0 = [n0, 0, 0]
    v1 = [-n0, 0, 0]
    v2 = [0, n0, 0]
    v3 = [0, -n0, 0]
    v4 = [0, 0, n0]
    v5 = [0, 0, -n0]
    points = [v0, v1, v2, v3, v4, v5]
    faces = [
        [0, 4, 2],
        [2, 4, 1],
        [1, 4, 3],
        [3, 4, 0],
        [2, 5, 0],
        [1, 5, 2],
        [3, 5, 1],
        [0, 5, 3],
    ]
    polyhedron(points, faces)
```

```python
def platonic_dodecahedron(n):
    phi = (1 + 5 ** 0.5) / 2
    fac = n * phi / 2
    n0 = phi * fac
    n1 = fac / phi
    v0 = [fac, fac, fac]
    v1 = [-fac, fac, fac]
    v2 = [fac, -fac, fac]
    v3 = [-fac, -fac, fac]
    v4 = [fac, fac, -fac]
    v5 = [-fac, fac, -fac]
    v6 = [fac, -fac, -fac]
    v7 = [-fac, -fac, -fac]
    v8 = [0, n0, n1]
    v9 = [0, -n0, n1]
    v10 = [0, n0, -n1]
    v11 = [0, -n0, -n1]
    v12 = [n1, 0, n0]
    v13 = [n1, 0, -n0]
    v14 = [-n1, 0, n0]
    v15 = [-n1, 0, -n0]
    v16 = [n0, n1, 0]
    v17 = [-n0, n1, 0]
    v18 = [n0, -n1, 0]
    v19 = [-n0, -n1, 0]
```

```
points = [
    v0,
    v1,
    v2,
    v3,
    v4,
    v5,
    v6,
    v7,
    v8,
    v9,
    v10,
    v11,
    v12,
    v13,
    v14,
    v15,
    v16,
    v17,
    v18,
    v19,
]
f1 = [0, 16, 18, 2, 12]
f2 = [14, 12, 2, 9, 3]
f3 = [19, 3, 9, 11, 7]
f4 = [15, 7, 11, 6, 13]
f5 = [4, 13, 6, 18, 16]
f6 = [2, 18, 6, 11, 9]
f7 = [12, 14, 1, 8, 0]
f8 = [3, 19, 17, 1, 14]
f9 = [7, 15, 5, 17, 19]
f10 = [13, 4, 10, 5, 15]
f11 = [16, 0, 8, 10, 4]
f12 = [1, 17, 5, 10, 8]
faces = [f1, f2, f3, f4, f5, f6,
         f7, f8, f9, f10, f11, f12]
polyhedron(points, faces)
```

```python
def platonic_icosahedron(n):
    phi = (1 + 5 ** 0.5) / 2
    fac = n / 2
    n0 = phi * fac
    v0 = [0, fac, n0]
    v1 = [0, fac, -n0]
    v2 = [0, -fac, n0]
    v3 = [0, -fac, -n0]
    v4 = [fac, n0, 0]
    v5 = [fac, -n0, 0]
    v6 = [-fac, n0, 0]
    v7 = [-fac, -n0, 0]
    v8 = [n0, 0, fac]
    v9 = [-n0, 0, fac]
    v10 = [n0, 0, -fac]
    v11 = [-n0, 0, -fac]
    points = [v0, v1, v2, v3, v4, v5, v6,
              v7, v8, v9, v10, v11]
    f1 = [2, 9, 0]
    f2 = [2, 7, 9]
    f3 = [2, 5, 7]
    f4 = [2, 8, 5]
    f5 = [2, 0, 8]
    f6 = [1, 4, 6]
    f7 = [1, 6, 11]
    f8 = [1, 11, 3]
    f9 = [1, 3, 10]
    f10 = [1, 10, 4]
    f11 = [0, 6, 4]
    f12 = [6, 0, 9]
    f13 = [9, 11, 6]
    f14 = [11, 9, 7]
    f15 = [7, 3, 11]
    f16 = [3, 7, 5]
    f17 = [5, 10, 3]
    f18 = [10, 8, 5]
    f19 = [8, 4, 10]
    f20 = [4, 8, 0]
```

```python
    faces = [
        f1,
        f2,
        f3,
        f4,
        f5,
        f6,
        f7,
        f8,
        f9,
        f10,
        f11,
        f12,
        f13,
        f14,
        f15,
        f16,
        f17,
        f18,
        f19,
        f20,
    ]
    polyhedron(points, faces)


# Local utility functions


def _radians(degrees):
    return degrees * math.pi / 180


def _cart(radius, angle):
    rad = _radians(angle)
    x = radius * math.cos(rad)
    y = radius * math.sin(rad)
    return [x, y]
```

About the Author
John Clark Craig

John Clark Craig has authored many books on programming topics, mostly covering versions of BASIC and Visual Basic languages as they evolved over time.

Today his focus is on Python, the world's most popular and easy-to-learn language, suitable for introducing young people to programming for the first time, yet powerful enough for the most challenging engineering, web design, gaming, robotic and machine learning ... actually ALL of today's hot programming areas.

In addition to writing books, John's software projects have controlled and monitored huge solar energy projects, helped wind engineers design better towers for wind turbines, monitored natural gas and oil projects in Alaska, helped train athletes for the U.S. Olympics teams, aided in the design of artificial knee replacement parts, provided a Python library for easier 3D design using OpenSCAD, and even provided tools for scientifically based research into the UFO phenomenon.

John lives in Colorado, today helping his wife with software tools that help her to help homeowners save lots of money by installing solar panels on rooftops. (See Solar-Proud.com)

Check out all his titles to show you just how passionate John is about the Python programming language and how it can be used in so many diverse ways to help others make the world a better place, one programming line at a time.

Other books by John Clark Craig

For a complete list of all the books
by John Clark Craig -
visit his website at

JohnClarkCraig.com

OpenSCAD Cookbook
OpenSCAD Recipes for learning 3D modeling -
ISBN: 1790273919

Python for 3D Printing
Using Python to enhance the power of OpenSCAD for
3D modeling -
ISBN: 1696881943

John Clark Craig

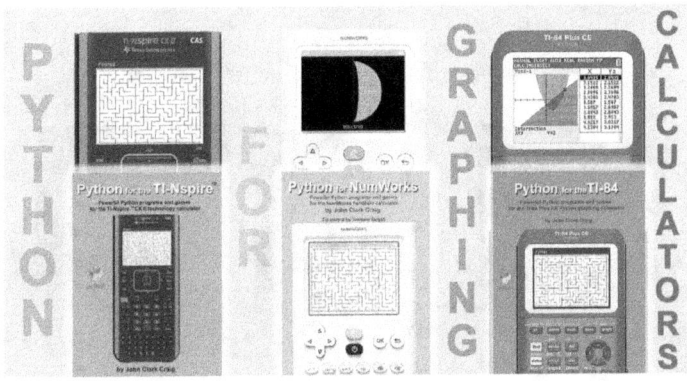

Python for NumWorks
Powerful Python programs and games for the NumWorks handheld calculator
ISBN: 979-8558337716

Python for the TI-Nspire™
Powerful Python programs and games for the TI-Nspire CX II technology calculator
ISBN: 979-8463835772

Python for the TI-84
Powerful Python programs and games for the TI-84 Plus CE Graphing Calculator
ISBN: 979-8476394686

VB-2012 - Random Numbers
introducing one of the best psuedorandom number generators
ASIN: B0075RJ42G

VB.NET - Sun Position
High accuracy solar position algorithms - a resource for programmers and solar energy engineers
ASIN: B005AJ93F4

VB-2012 - Strings
programming examples of enhanced string processing
ASIN: B004G095MO

John's books from other publishers

Visual Basic 2005 Cookbook
John Clark Craig & Tim Patrick
O'Reilly

Microsoft Visual Basic: Developer's Workshop
John Clark Craig & Jeff Webb
Microsoft Press

For a complete list of all the books
by John Clark Craig -
visit his website at

JohnClarkCraig.com

To Contact John

for code downloads
to guest speak
or to consult:

email him at
john@craigware.com

www.ingramcontent.com/pod-product-compliance
Lightning Source LLC
Chambersburg PA
CBHW070626220526
45466CB00001B/107